RSPB

RSPB POCKET GUIDE TO
BRITISH BIRDS
SECOND EDITION

Simon Harrap

Illustrated by Dave Nurney

B L O O M S B U R Y
LONDON · BERLIN · NEW YORK · SYDNEY

Second edition published 2012 by Bloomsbury Publishing Plc,
50 Bedford Square, London, WC1B 3DP

Copyright © 2007, 2012 text by Simon Harrap
Copyright © 2007, 2012 inside illustrations by Dave Nurney
Copyright © 2012 cover illustration by Stephen Message

ISBN (print): 978-1-4081-7456-2

A CIP catalogue record for this book is available from the British Library.

Commissioning editor: Julie Bailey
Project editor: Jasmine Parker
Designed by Rod Teasdale

Printed in China by C&C Offset Printing Co., Ltd.

10 9 8 7 6 5 4 3 2 1

MIX
Paper
FSC FSC® C008047

CONTENTS

a million voices for nature

The RSPB speaks out for birds and wildlife,
tackling the problems that threaten our environment.
Nature is amazing – help us keep it that way.

If you would like to know more about The RSPB,
visit the website at www.rspb.org.uk
or write to:
The RSPB
The Lodge
Sandy
Bedfordshire
SG19 2DL
Tel: 01767 680551

AUTHOR ACKNOWLEDGEMENTS
Thanks to Dr David Leech for reading through the entire text and
making many useful suggestions, to Nigel Redman of Bloomsbury
and his wife Cheryle Sifontes for their enthusiastic support and to
Julie Bailey and Jasmine Parker of Bloomsbury for their expert
guidance.

IMAGE CREDITS
p.13, aleks.k/Shutterstock; p.14, Christopher Elwell/Shutterstock;
p.16, Gail Johnson/Shutterstock; p.17, stocker1970/Shutterstock;
p.18, Stephen Meese/Shutterstock.

INTRODUCTION

Birds are undoubtedly the most popular aspect of Britain's wildlife. The membership of the RSPB, standing at well over a million, dwarfs other conservation organisations. But, while many people are interested in birds and concerned for their welfare, relatively few are able to identify more than a few distinctive species. How can this be, when there are dozens of field guides and other bird books? Taking a closer look, I realised that while the expert birder is now very well catered for with several excellent field guides, the beginner is rather poorly-served. It thus seemed that there was a need for one more book, specifically written and designed for those starting out on their birdwatching journey and eager to identify the birds that they see around them.

This field guide has therefore been designed with the following in mind:

- It should include all the species of bird likely to be encountered in the garden, on country walks or on visits to the seaside or bird reserves, but no more.

- It should be specifically designed for use in Britain and Ireland.

- It should be written, as far as possible, in non-technical terms.

- It should be written, designed and illustrated to the highest standard.

- It should be affordable.

- It should be genuinely pocket-sized.

HOW TO USE THIS BOOK

Each species has its own entry (a few very similar species are treated together), and each entry is written to a standard format. The English name is followed by a measurement giving the approximate total length of the bird from the tip of the bill to the tip of the tail in centimetres. The title line also includes the scientific name, which is always made up of two words and written in italics. Scientific names are based on Latin or ancient Greek or are Latinised versions of the names of people or places. The system of scientific names is international and subject to strict rules, and is designed to give every species of animal and plant a name that can be understood everywhere in the world, whatever the local language.

After a brief introduction the main body of the text is divided up under the following headings: Description, Population, Habitat, Voice and Confusion Species. These headings are self-explanatory, but under Confusion Species a few additional species are mentioned that are not illustrated in the *Pocket Guide* but which may cause confusion.

KEY TO MAPS
The maps in this book give an approximate indication of the distribution of each species at different seasons of the year.

Green: resident, areas where species may be seen throughout the year and where they breed

Yellow: summer visitor, areas where the species may be seen in summer and usually breed

Blue: winter visitor, areas where species spend the winter, but do not breed

Pink: passage migrant, areas that species visit at times of migration – generally spring and autumn

For distribution at sea, colours are restricted to areas where birds will be visible to observers and therefore only inshore waters have been mapped.

TOPOGRAPHY

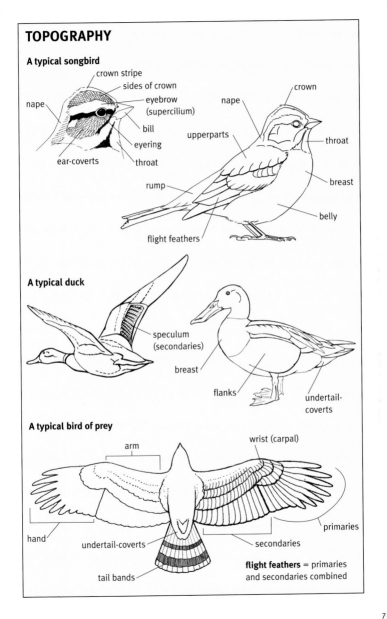

A typical songbird

crown stripe
sides of crown
eyebrow (supercilium)
nape
bill
eyering
ear-coverts
throat

crown
nape
upperparts
throat
breast
rump
belly
flight feathers

A typical duck

speculum (secondaries)
breast
flanks
undertail-coverts

A typical bird of prey

arm
wrist (carpal)
hand
primaries
undertail-coverts
secondaries
tail bands

flight feathers = primaries and secondaries combined

A typical gull

outer wing

mirror

inner wing

primaries

secondaries

rump

uppertail-coverts

tail

crown

nape

forecrown

mantle

bill

scapulars

breast

wing coverts

wingtips

tail

legs

belly

flanks

feet

A typical wader

crown

mantle

bill

scapulars

flight feathers

breast

belly

wing coverts

IDENTIFYING BIRDS

Identifying birds is fun, mostly rather easy and very rewarding. The secret of success is to look carefully and to listen.

LOOKING

In the garden many birds can be seen well enough to be identified without the use of any optical aids, but in more open country or on the coast a pair or binoculars is almost essential. Indeed, apart from a field guide, a pen and a notebook, binoculars are the only equipment needed to be a birdwatcher.

BINOCULARS

Binoculars are available at a huge range of prices, from a few pounds to a thousand pounds. In general, you get what you pay for, but there are a few points worth remembering.

1. Binoculars are specified by two numbers, e.g. 10 x 42, 8 x 40. The first figure gives the magnification, the second the diameter in mm of the objective lens – the larger lens, not the eyepiece that you look into. Never buy binoculars with a magnification of more than ten! You will not be able to hold them steady enough and their optical quality is probably poor; conversely, a magnification of less than seven is not likely to be of much use. In general, the bigger the objective lens, the more light the binoculars let in and the brighter the image, but modern high-tech coatings have made smaller objective lenses more useful, and excellent (if expensive) miniature binoculars with objective lenses as small as 20 mm diameter are available.

2. Ideally, binoculars should be sealed and fully waterproof. In our climate it is all too easy to get moisture inside binoculars and if permanently 'steamed-up' an expensive repair will be necessary.

3. Always try before you buy. The size, shape and handling of binoculars are important considerations and are very personal, so hands-on experience is essential. Any good dealer will allow you plenty of time to look at and handle a range of models.

4. Second-hand binoculars are well worth considering. As with cars, many birdwatchers trade-in perfectly good binoculars for the latest model and thus premier marques can be had at bargain prices.

If binoculars are essential, a telescope and a tripod are a luxury that you may wish to invest in as your interest develops, allowing you to get really good looks at distant birds. A telescope is particularly useful if you spend time looking at wildfowl and waders. It is essential to purchase a 'scope designed for birdwatching, with either a fixed magnification eyepiece of 20x to 40x or a zoom somewhere in the range 20–60x. (Note that astronomical telescopes produce an upside-down image!) As with binoculars, try before you buy and remember that you get what you pay for, although second-hand is again a good bet.

LISTENING

Birds use a variety of calls and many also have a song. Although these are nowhere near as complex and sophisticated as human speech, they serve much the same purposes. Calls are generally short, simple sounds that communicate specific things – most birds have contact calls, alarm calls (with different alarms for different types of predators) and begging calls. Songs are usually rather more complex and mostly given by males, they serve to defend a territory and attract a mate.

Songs and calls play an important part in the life of birds, and are extremely useful to the birdwatcher too. In the first place they help us to find the birds and they also help us to identify birds. Indeed, for some species songs and calls are by far the best clue to their identity. Willow Warbler and Chiffchaff look very similar, but their songs are totally different; Marsh and Willow Tit are even more similar in appearance but, used with care, their calls are by far the best way to separate them.

Despite their usefulness, however, many birdwatchers struggle to recognise songs and calls. One problem is that it is very difficult to give written descriptions of calls and even more so songs. One way around this is to listen to recordings, and selections of superb collections are available on CD, although for many people this is still confusing. But, if you can recognise the music of different composers or the tunes produced by different bands, you can recognise bird sounds.

My advice is to start by learning the vocalisations of the birds that you can hear every day in the garden or at work. You may need to put some time into this, listening carefully, working out what is distinctive to your ears and identifying the songster, but who could complain about time spent in the garden listening to the birds? In some species the song is stereotyped, and sounds more or less the same whichever individual gives it or wherever you are. Examples include Chaffinch, and these songs are relatively easy to learn. In other species the song varies between individuals and often changes over time too, making it harder to learn. Examples include Blackbird, Song Thrush and Robin. But, just as a particular pop group has a 'sound', each species has its own sound and it is not really very difficult to learn to recognise their song. Once you have established a base-line with the common, familiar species that you hear regularly, try to extend your knowledge when you are out and about by picking out the sounds that you do not recognise and then tracking down and identifying the culprit.

FIELD NOTES

Keeping a notebook is an excellent way of getting more out of your birding. Why take notes? Because however good your memory, a notebook is the only reliable way to record facts about the birds that you see. How else would you know the date of your first Swallow in 2001, or when the first brood of Blackbirds fledged in your garden last year? There are many different systems of note-taking. Some people have just one notebook and fill it in as they go along. Others prefer to keep a rough field notebook and transfer the information to a neat, permanent notebook when they get home (or, increasingly, to their computer).

A typical notebook entry would include the date, the place and perhaps also the time spent there and who you were with; I have also always included details of the weather and the route taken. As for the birds, you may want to list every species seen or just the more interesting ones, and you may well want to record the numbers seen of each species, whether it be accurate counts, estimates or more likely a mixture of the two. You do not need to confine your note-taking to formal days out 'birdwatching', however; you can also jot down details of birds in the garden, seen on the way to work or indeed anywhere. Whenever you notice a bird, it is noteworthy.

As well as details of the birds that you recognise, a field notebook is essential when you see an unfamiliar bird, as you can take notes on what it looks like. Such field notes are invaluable, because the very process of writing things down helps us to look more carefully at the bird – e.g. what colour are the legs, where exactly is that white stripe on the head? Even better, do a sketch and annotate it with detail of the plumage, bare part coloration, size and shape. You do not need any artistic ability to make field sketches. A circle for the head, a bigger one for the body and a couple of stick legs will do the job! Drawing a sketch really focuses your attention on the whole bird and will often prompt you to record details that will be useful later when you come to identify the bird. And, importantly, making notes on the spot provides objective evidence and prevents wishful thinking when you get home. You will find that time spent note-taking is well spent, not only because your notes will build up into a valuable reference, but also because reading through them is a great way to re-live your birding experiences.

Modern digital technology enables birders to capture photos, videos and sound recordings of birds, which can be used to supplement more traditional field notes in order to check up on identifications or seek expert help. Digital cameras or mobile phones can be used to take quick reference photographs or short films, and used in conjunction with a telescope ('digi-scoping') can produce great results. Songs and calls can be recorded on digital recorders or smart phones, although a plug-in directional microphone will significantly improve the quality of the recordings. There are also apps rapidly coming onto the market that help bird identification in the field and the recording of bird data.

IMPROVING YOUR FIELDCRAFT

A few simple techniques will help you to get more out of your birding, by seeing more birds and seeing them better.

- Get out early and late. Especially in hot summer weather birds are much more active early and late in the day. The light is usually much better too, allowing you to see colours more accurately and avoiding heat-haze – the shimmer produced by hot air rising from the ground can render even the most expensive optics worthless!

- Use the light. Plan your route, or your approach to an interesting bird, to have the sun behind you or to the side. It is astonishing how much difference this can make, and poor views of a near-silhouette as you squint into the sun can be transformed into clear, well-lit views, with every colour clearly visible, by getting this right.

- Avoid bright, gaudy clothing. Even for birds that are not shy, vivid colours draw their attention unnecessarily.

- Use your ears and be quiet; it is hard to hear birds if you are making a lot of noise. Try to avoid clothing that 'swishes' as you walk, an unfortunate characteristic of many recent 'technical' fabrics used in outdoor clothing.

- Avoid 'sky-lining'. If you appear on the skyline you will be painfully obvious and this may lead shy birds such as wildfowl and waders to depart post-haste. Walking at the base of a bank rather than along the top, or choosing a route where a hedge forms a backdrop, can get you closer to the birds.

- If you are trying to get closer to a bird, do not walk directly towards it, rather take a roundabout route. If the bird senses it is the exclusive subject of your attention it will feel more threatened. Similarly, avoid sudden movements. It is not necessary to freeze, but slow, gentle movements are far less alarming.

- Wear suitable clothing. It is hard to enjoy birds if you are cold and wet or hot and thirsty. Even in the summer, conditions can change quickly, so the advice is to wear several layers that can be put on or taken off as appropriate. In hot sunshine a hat can make all the difference, and don't forget sunblock and water; especially on the coast, it is all too easy to get sunburnt and dehydrated.

HABITATS

GARDENS

Garden birds are the jumping-off point in their birdwatching career for most people and, for many, watching birds in the garden remains the core of their birding and gives great pleasure. Indeed, typical garden birds such as Wren, Blackbird, Song Thrush, Robin, Dunnock and Great and Blue Tits are amongst the nation's favourites. The size of the garden and variety of trees and shrubs present clearly affect the variety of species that can be seen, but adding a pond and putting out food are obvious ways of increasing your garden's attractiveness to birds. It is also worth remembering that birds do not take any notice of property boundaries and it is the *total* of the available habitats in your area and the proximity of woods, farmland and other habitats that determines what you might see. This is good news, because whatever the size of your garden, you can attract birds.

PARKS AND CEMETERIES

Within towns and cities the more extensive areas of mature trees and shrubs found in many parks and cemeteries provide a home for woodland birds such as pigeons, Tawny Owl and Great Spotted Woodpecker, and park lakes often attract a few wild ducks. Large areas of grassland, such as playing fields, may attract wintering Fieldfares and Redwings.

Above Parks are a good starting point for birding, especially if there are trees and shrubs for birds to seek refuge in. Parks are home to a range of species including Blue Tit, Wren, Robin, Great Spotted Woodpecker and sometimes Tawny Owl.

Above Farmland is home to a wide variety of birds, especially where there is a mixture of habitats, such as mature hedges, copses and ponds, and it is being managed in a wildlife-friendly way.

FARMLAND

Farmland comprises a mosaic of habits, including arable crops, pastures, hedgerows and copses and, quite simply, the greater the variety of habitats, the greater the variety of birds. The huge, prairie-like intensively farmed arable fields of parts of eastern England may hold little more than visiting Wood Pigeons and Rooks, whereas the intricate mix of habitats in some other regions may have a variety of finches and warblers breeding in the hedgerows and, in winter, flocks of Lapwings, gulls, thrushes, finches and buntings. Farmland is a habitat often neglected by birdwatchers in favour of more 'exciting' habitats around inland waters or the coast, but it is always worth exploring your local area.

WOODLAND

Woodland comes in great variety, but all woodland in Britain is heavily influenced by man's activities. And, although birds are not as fussy as plants, fungi or insects, the type of woodland affects the birds that can be seen. At its worst, 'woodland' consists of plantations of conifers. In their young stages these are low and scrubby and attract a variety of birds, but as they mature the sombre ranks of 'trees on parade' may hold little more than Wood Pigeons, Goldcrests and Coal Tits. At best, woodland consists of a mixture of native trees (which include Scots Pine in Scotland) and, most importantly for birds, has a varied structure, with trees of all ages, from saplings to dying and dead trees and with broad, sunny rides, glades and clearings. Such woods are a delight, especially in spring, with Turtle Dove, a variety of warblers including Garden Warbler and a variety of tits including Marsh Tit. Nightingale is a speciality in the south-east, especially in actively-managed coppice, while the oak woodlands of the north and west hold Redstart, Wood Warbler and Pied Flycatcher.

Above Deciduous woodland is at its best in early spring when most birds are in song and easier to locate; by high summer the dense canopy makes birds harder to see. In winter bare trees again offer good birding, with the resident woodpeckers often obvious.

Above Moorland is home to Golden Plover, Dunlin, Merlin, Hen Harrier and also Red Grouse, which is virtually confined to heather moorland.

HEATHLAND
Found on poor, sandy soils in the lowlands, heathland is dominated by shrubs such as gorse, Broom and heathers, often mixed with grasses and Bracken and with scattered pines and birches. Typical birds include Green Woodpecker, Tree Pipit, Willow Warbler, Linnet and Yellowhammer, while specialities include Nightjar and Stonechat.

MOORLAND
The upland counterpart of heathland, but dominated by heathers and rough grasses and often including more extensive wet, boggy areas. Typical birds include Curlew, Meadow Pipit and Skylark, while specialities include Red Grouse, Hen Harrier and Golden Plover.

RIVERS AND STREAMS
Lowlands rivers and streams tend to be slow-flowing and are often bordered by aquatic vegetation. Typical birds include Mute Swan and Kingfisher. In the uplands the steeper gradients means that the waters are faster-flowing, and typical birds include Common Sandpiper, Grey Wagtail and Dipper.

LAKES, RESERVOIRS AND GRAVEL PITS

In many areas these provide the most exciting and varied habitats for birdwatchers. Size, depth and the level of disturbance all affect the number and variety of birds that may be seen. Grebes, Coots, Moorhens and a wide selection of wildfowl can be seen on and around the water, as can Herons and Kingfishers. Gulls occur throughout the year, but in winter larger numbers gather to roost overnight on some inland waters, while in summer Common Terns may breed. Waterside vegetation holds its own special birds, with stands of Common Reed being particularly attractive, and in summer Reed and Sedge Warblers and Reed Bunting may be found. The area around most reservoirs and gravel pits also contains attractive areas of scrub with a wide variety of breeding birds, such as Turtle Dove, Garden Warbler and Common and Lesser Whitethroats. For many birdwatchers, however, the migration periods of May and late July–September can be the most exciting. A wider variety of terns may occur and, if water levels drop to expose a muddy shoreline (especially likely in the autumn), several species of waders can be found.

BEACHES AND DUNES

Most beaches are too disturbed in summer to attract many birds, but in some favoured nature reserves there may be breeding Oystercatchers, Ringed Plovers and Little Terns. Extensive areas of dunes hold more breeding birds, including common species such as Skylark and Meadow Pipit as well as the specialists: gulls, terns and, in the north, Eider. In the winter, when the tourists are absent, dunes may be deserted, but beaches hold a variety of gulls and waders, with Sanderling and Snow Buntings the specialities.

Below Sandy and shingle beaches are all too often heavily disturbed for much of the year but, when protected, will hold breeding Ringed Plovers, Oystercatchers and perhaps also terns. Most beaches are deserted by holidaymakers in winter so this can be a good time to visit for a variety of waders and gulls, as well as auks and ducks offshore.

ESTUARIES, SALTMARSHES AND MUDFLATS

The rise and fall of the tide creates a range of habitats on gently shelving shores. In the area between the normal high- and low-water marks there is little vegetation, and mudflats are the norm. These may be muddy or sandy and can attract huge numbers of feeding wildfowl and waders, creating some of the great bird spectacles, especially in estuaries where there is a continual input of nutrients. Slightly higher up the shore, on areas that are only flooded by seawater on the highest tides, conditions are more conducive to the growth of vegetation, and saltmarsh develops, usually drained by an intricate network of deep muddy gutters. Fewer birds feed on the saltmarsh, typically with Black-headed Gulls, Redshanks and Greenshanks in the gutters, but they can hold breeding gulls and waders in summer and a variety of small seed-eaters, such as Linnet and Reed Bunting, in winter.

CLIFFS

Cliffs provide the home for some of Britain's most important birds – breeding seabirds. Where the type and structure of the rocks provide suitable ledges, mostly in the north and west, huge colonies can be found, and cliffs also provide a home for Peregrine, Rock Dove, Raven and Chough. Even in the winter months, some seabirds may be present, joined by wintering Purple Sandpipers and Turnstones.

Below RSPB Bempton Cliffs Reserve, in East Yorkshire, provides shelter for around 200,000 nesting seabirds, including Guillemots, Gannets, Razorbills and Puffins.

GLOSSARY

Buff: a yellowish-beige colour.

Eclipse plumage: the drab, female-like plumage worn by male ducks for a few weeks in summer during the period when they are moulting their wing feathers and are flightless or nearly so.

Eyebrow: a pale line on the side of the head, extending from the bill back over and above the eye (more technically, the 'supercilium').

Immature plumage: the plumage worn by a young bird in the period between the juvenile plumage (q.v.) and the adult plumage. In some species, especially larger birds such as gulls and birds of prey, there may be several immature plumages, each one progressively more like the adult.

Juvenile plumage: the plumage worn by a young bird when it first leaves the nest, often rather dull and cryptically coloured. In most cases this is quickly replaced, either by an adult-type plumage (e.g. a juvenile Robin has a spotted brown breast but rapidly moults and replaces this with a red breast), or by an immature plumage.

Leaf warbler: Willow Warbler, Wood Warbler and Chiffchaff, all members of the genus *Phylloscopus*, so called because of their green and yellow, leaf-coloured plumage.

Lek: an early morning gathering where several males display and attendant females select the most impressive performer to mate with.

Machair: a rare habitat found in the Western Isles of Scotland on low-lying areas near the coast and comprising flower-rich grassland growing on soils made up of wind-blown sand mixed with peat.

Passage: another word for migration, the journey between summer and winter quarters.

Pelagic: living out at sea.

Primaries: the outer flight feathers.

Red List The official list compiled by government agencies of birds that are suffering declines and in danger of extinction in Britain.

Sea-watching: watching for birds passing the coast, often undertaken during stormy onshore winds and requiring a good telescope, warm clothing and much patience!

Secondaries: the inner flight feathers.

Speculum: the inner flight feathers of ducks, often metallic green or blue, or contrastingly patterned with black and white. Often visible on the closed wing as a small oblong patch, but most obvious in flight.

Tubenose: a term for Fulmar and several close relatives that have prominent tube-shaped nostrils on their bill, an adaptation allowing them to excrete excess salt.

COLOUR-CODING KEY

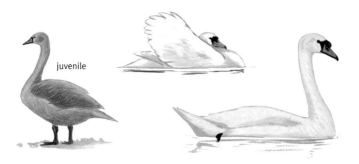

juvenile

One of Britain's largest birds and the most commonly encountered swan. Although the least vocal of swans, it is far from mute.

Heads of Juveniles

DESCRIPTION The classic swan, neck held straight or in graceful S-curve, often carries wings raised up over back. Bill orange-red with black base and prominent black knob on forehead. Sexes similar, although male (cob) has a larger knob and more richly coloured bill than female (pen). Juveniles grey-brown with flesh-grey bill lacking knob, becoming white over 1–2 years. Can be aggressive and capable of inflicting injury; angry swans should be given a wide berth.

Mute

Bewick's

Whooper

POPULATION Fairly common resident; numbers have increased in recent years, perhaps helped by a ban on lead fishing weights, which swans can ingest accidentally.

HABITAT Well-vegetated lakes, reservoirs, gravel pits, larger ponds and slow-flowing rivers, also sheltered coastal waters.

VOICE Various calls, including a threatening hiss. In flight the wings make a unique rhythmic 'chuffing' noise.

CONFUSION SPECIES Adults distinguished from Whooper and Bewick's Swans by upward-pointing tail, bill colour and pattern and, in flight, noise made by wings. Juvenile similar to juvenile Whooper Swan, although darker and browner, with black bill-base.

BEWICK'S SWAN · 115–127 CM · *Cygnus columbianus*
WHOOPER SWAN · 145–160 CM · *Cygnus cygnus*

Bewick's

Bewick's

Whooper

Whooper

Winter visitors to the British Isles, these elegant 'wild' swans have beautiful, bugling calls but are hard to separate unless seen well.

DESCRIPTION Bill pattern diagnostic: on Bewick's the yellow forms a rounded patch on bill-base; Whooper has more yellow, extending in a wedge towards bill-tip. Bewick's also more compact with shorter neck. Juveniles of both are greyish and show ghost of adult bill pattern.

POPULATION Locally common winter visitors (Oct–Mar). Bewick's breeds on Siberian tundra and winters E England (especially Ouse and Nene Washes, Cambridgeshire), around Severn Estuary (including Slimbridge, Gloucestershire), in Lancashire (notably Martin Mere) and in S Ireland. Whooper breeds in Iceland and is widespread in Scotland and Ireland and more scattered in N England and East Anglia (especially Ouse Washes); has bred in Scotland.

Bewick's Swan

HABITAT Both forage on wet grassland and increasingly also on arable land, with Whooper Swan also on tidal mudflats, and both bathe, roost and loaf on freshwaters.

VOICE With practice the higher-pitched, more yelping voice of Bewick's can be distinguished.

CONFUSION SPECIES Both separated from Mute Swan by bill pattern, usually straighter neck and, in flight, bugling calls and absence of wing noise.

Whooper Swan

adult European
Whitefront

black bars
on belly

*The classic 'wild goose', the most famous haunt of this
winter visitor is Slimbridge in Gloucestershire, but it
occurs at scattered sites elsewhere.*

adult Greenland
Whitefront

1st-winter
European
Whitefront

DESCRIPTION A 'grey goose' with orange legs. Adults
have white blaze at base of bill and dark bars on belly.
Juvenile more uniform, the white blaze starting to
develop late in the first winter. Two populations are
identifiable: European Whitefronts have a reddish-pink
bill; while Greenland birds, subspecies *flavirostris* are
larger and darker, with a bulkier orange-yellow bill
(pinkish towards tip).

POPULATION Locally common winter visitor. Greenland
Whitefronts occur Oct–late Apr in Ireland (mostly on
Wexford Slobs), Scotland (SW, Hebrides, especially Islay,
and Caithness) and W Wales, and are usually seen in
scattered groups. European Whitefronts are present
Oct–Mar, although the bulk arrive in Dec, and occur in
large flocks by the Severn Estuary, patchily elsewhere in
S England and East Anglia.

HABITAT European birds winter on wet pastures by
rivers and estuaries. Greenland Whitefront is more
adaptable, also using arable fields and bogs.

VOICE Flocks give a slightly muted cackling chorus.

CONFUSION SPECIES As with all 'grey geese', hard to
identify unless close enough to see bare part coloration
and plumage detail.

BEAN GOOSE • 66–84 CM • *Anser fabalis*

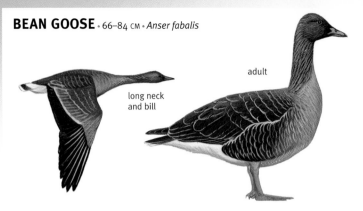

adult

long neck
and bill

Taiga Bean

Tundra Bean

A scarce winter visitor from Scandinavia, with just two regular wintering flocks, in the Yare Valley in Norfolk and Stirlingshire in central Scotland.

DESCRIPTION A large goose, around size of Greylag, the darkest of the 'grey' geese, especially around head and neck. Legs orange, bill dark with variable amounts of orange and sometimes a narrow whitish band around base. In flight rather dark, lacking grey on the upper- or underwing.

POPULATION Around 300 winter in central Scotland, early Sep–late Feb, but tend to be very elusive, while 150 are present in Norfolk mid Nov–early Feb. These two regular wintering flocks involve 'Taiga Bean Geese' (subspecies *fabalis*); 'Tundra Bean Goose' (subspecies *rossicus*, which is slightly smaller, and usually has only limited orange near tip of bill- much like Pink-footed Goose) is scarce and erratic visitor in small numbers to SE England, especially in hard weather.

HABITAT Wet grassland in Norfolk, fields and moorland in Scotland.

VOICE More honking than Pink-footed Goose, and lacking its squeaky *wink-wink* notes.

CONFUSION SPECIES Pink-footed Goose is smaller, shorter-necked and smaller-billed, with blue-grey upperwings and pink feet. Bean Geese with white around bill base could be confused with White-fronted Goose.

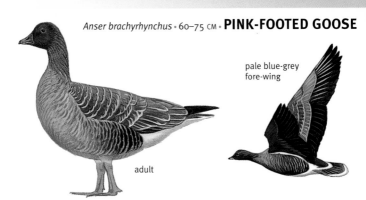

Anser brachyrhynchus • 60–75 CM • **PINK-FOOTED GOOSE**

pale blue-grey fore-wing

adult

In winter Britain holds most of the world population of this small 'grey goose' that breeds in Iceland and eastern Greenland.

adult

bill short and stubby

DESCRIPTION Relatively short-necked and compact. Legs pink, although colour often hard to see. Bill short and triangular, mostly dark, with small pink band across outer half. Dark head and upper neck contrasts with pale lower neck and body. In flight note rather short neck and pale upperparts, with most of upperwing pale blue-grey. Sexes similar. Juvenile may be duller and more uniform.

juvenile

POPULATION Common winter visitor (Sept–Apr) to E Scotland, Solway Firth, NW England (mainly Lancashire), Lincolnshire and Norfolk. Has increased from around 30,000 in 1950 to 225,000 by 2005 (of which up to 150,000 in Norfolk). Rare away from regular wintering areas.

HABITAT Pastures and arable fields. Usually found in large flocks, which can fly long distances to roost on large inland waters or coastal sandbars, but increasingly also roosts on grazing marshes.

VOICE Flocks give a cackling chorus, generally higher-pitched and more musical than other geese, in which it is possible to pick out clear, high *wink-wink* notes.

CONFUSION SPECIES Hard to separate from other 'grey geese' unless seen well, but location often a good clue.

GREYLAG GOOSE · 75–90 CM · *Anser anser*

pale grey
fore-wing

The only goose to have genuine wild breeding populations in Britain, but in the south most birds are from introduced stock and may be rather tame.

DESCRIPTION Largest 'grey goose', with bulky, pinkish-orange bill and pinkish legs. In flight has dark upperparts contrasting with large areas of pale grey on upperwing, also pale grey rump and pale underwing-coverts contrasting with dark flight feathers. Sexes similar. Juvenile as adult.

POPULATION Truly wild birds are resident in the Outer Hebrides and N Scotland, and these are supplemented in winter by the entire Icelandic population (around 100,000 birds), which winters in Scotland and NE England. Further south, the expanding resident populations probably derive exclusively from human introductions.

heavy, pinkish-orange bill

HABITAT Breeds around freshwaters, often nesting on islands, and feeds on nearby farmland, both arable and pasture, as well as amenity grasslands. Returns to lakes and reservoirs to roost.

VOICE A harsh cackling similar to a farmyard goose (this is the ancestor of most domestic breeds).

CONFUSION SPECIES Only 'grey goose' likely to be seen in summer; in winter, good views of bill and leg colour required, although distinctively pale in flight and larger and bulkier than other 'grey geese'.

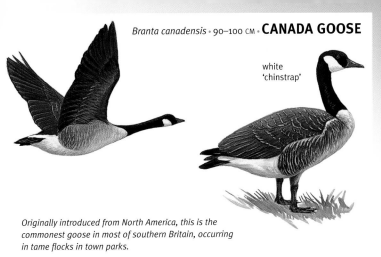

Branta canadensis • 90–100 CM • **CANADA GOOSE**

white 'chinstrap'

Originally introduced from North America, this is the commonest goose in most of southern Britain, occurring in tame flocks in town parks.

DESCRIPTION Large. Brownish body, pale breast, black neck and white 'chinstrap' distinctive. Sexes similar. Juvenile slightly duller, with head and neck more brownish-black and cheeks washed brownish.

POPULATION Common resident. Introduced around 300 years ago to the estates of wealthy landowners, it was breeding freely by the late 18th century, although still largely a bird of parks and ornamental lakes as recently as the 1940s. Has since undergone a population explosion and spread widely, so much so that it is considered a pest in some areas. Some birds from central and N England migrate to Beauly Firth in Scotland to moult. Genuinely wild Canada Geese, vagrants from N America, are occasionally found with flocks of wild geese.

HABITAT Reservoirs, gravel pits, rivers, canals and ornamental park lakes. Feeds on grassland.

VOICE Flocks give a loud honking with clear, well-spaced notes.

CONFUSION SPECIES Separated from Barnacle Goose by brown rather than grey plumage, with pale breast, and white 'chinstrap', rather than black breast and rounded white patch at base of bill.

BARNACLE GOOSE · 58–70 CM · *Branta leucopsis*

The large flocks of Barnacle Geese wintering on the Solway Firth and Islay form a great wildlife spectacle, but further south genuine wild birds are rare.

white face-patch at base of bill

DESCRIPTION Upperparts grey, black of neck extends onto breast with circular white face-patch at base of bill. In flight note light grey upperwing. Sexes similar. Juvenile as adult but has subtly mottled rather than barred flanks.

POPULATION Winter visitor (Oct–Apr). Birds breeding in E Greenland winter in N and W Scotland, south to Islay, and on islands off N and W coasts of Ireland (with a few on other Irish coasts and on Skomer Island, Pembrokeshire). Birds breeding in Spitsbergen winter exclusively on the Solway Forth (Caerlaverock in Dumfries & Galloway and Rockcliffe in Cumbria). Singletons and small flocks seen further south are usually from feral populations but, especially in hard weather, wintering wild birds from the Netherlands may visit coasts of SE England.

HABITAT Winters on saltmarsh and pastures close to the sea and on small grassy offshore islands.

VOICE A gruff honking.

CONFUSION SPECIES Very distinctive if seen well. White face distinguishes it from Brent Goose, while Canada Goose is much larger and brownish rather than grey, with pale breast and white 'chinstrap'.

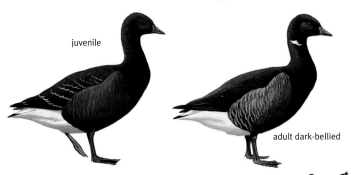

juvenile

adult dark-bellied

The British Isles are the winter home of two different forms of this small goose, Dark-bellied and Pale-bellied, and hold nearly half the world population of the former.

DESCRIPTION Smallest goose, overall rather dark with contrasting white rear end. Sexes similar. Adults have small white crescent on sides of neck and uniform upperparts. Immatures have fine whitish bars on upperparts and only develop white crescent late in winter. Two distinct forms occur. Dark-bellied Brent has dark grey belly and flanks with limited pale area on sides of body; Pale-bellied Brent (subspecies *hrota*) has much paler underparts, contrasting with blackish breast, and more brownish upperparts (NB flanks initially dusky in immatures). Flies in dense, disorderly flocks. Often sits on sea, and will up-end like a duck.

POPULATION Winter visitor, present Sept–Apr. Dark-bellied Brent breeds in arctic Siberia and occurs along the E and S coasts of England, Pale-bellied Brent, subspecies hrota breeds in Spitsbergen and winters in NE England (mainly Lindisfarne), and also breeds in Greenland and Canada, wintering in Ireland. Both forms are rare inland.

HABITAT Mudflats and coastal grassland.

VOICE Flocks give a chorus of low, throaty *krrrs*.

adult pale-bellied

CONFUSION SPECIES Barnacle Goose is larger, with crisp grey upperparts and white face.

EGYPTIAN GOOSE • 63–73 CM • *Alopochen aegyptiaca*

in flight shows striking black and white wings

Related to shelducks rather than geese and originally from Africa, this species was introduced to England in the late 17th century.

juvenile

DESCRIPTION Slightly larger than Shelduck. Pinkish-brown overall, face white with dark patch around eye, giving rather startled expression. Bill dull pinkish-red with dark cutting edges and tip, legs dull pinkish-red. Sexes similar. Juvenile duller, with darker and less contrasting head, no dark belly patch and greyish legs. In flight white wing-coverts contrast with dark flight feathers on both upper- and underwing. Found in pairs in spring and forms large flocks in late summer. May be seen perching in trees.

POPULATION Fairly common in Norfolk and N Suffolk, it is slowly increasing and spreading and is increasingly recorded elsewhere in S England.

HABITAT Parkland and farmland, especially around lakes and rivers. Breeds in tree holes, occasionally in an old crow's nest or on the ground.

VOICE A variety of harsh honking and quacking notes.

CONFUSION SPECIES Ruddy Shelduck *Tadorna ferruginea* sometimes escapes from captivity and has a similar wing pattern, but is rufous overall with black bill and legs.

Tadorna tadorna • 58–67 CM • **SHELDUCK**

female

male

Strikingly piebald and halfway between a duck and a goose, Shelducks are commonly seen in pairs and small parties on sheltered estuaries and coastal lagoons.

juvenile

DESCRIPTION Size as small goose. Drake has bright red bill with prominent knob, especially in breeding season. Female's bill often duller, with smaller knob (most obvious in direct comparison), while chestnut and black markings on breast and belly are narrower and less well-defined. Juvenile very distinct, lacking chestnut breast-band and with crown and nape brownish-grey, contrasting with white face, throat and foreneck.

POPULATION Common resident, although in Jul–Oct most adults fly to Heligoland Bight in Germany to moult. Some birds remain in Britain, gathering to moult at Bridgwater Bay and large estuaries such as the Wash, Humber and Forth. Numbers are supplemented in winter by Continental birds.

HABITAT Estuaries, with small numbers inland around lakes, reservoirs and gravel pits. Nests in burrows, sometimes tree holes, often far from water.

VOICE Thin whistles, often given by male in flight whilst chasing a female, and a hoarse *k-k-k-k-krrr, krrr, krrr*.

CONFUSION SPECIES None. Juveniles could be mistaken for similarly white-faced Egyptian Goose, but are otherwise very different. See also Goosander and Red-breasted Merganser.

MANDARIN · 45 CM · *Aix galericulata*

male

female

adult

A spectacular duck, introduced to Britain in the early 20th century, since when it has slowly increased and spread outwards from its original stronghold in Surrey.

DESCRIPTION Male distinctive. Female grey-brown with shaggy crest, white ring around eye extending backwards into white line curving down side of head, white around base of bill, and white chin.

POPULATION Native to Russian Far East, E China and Japan, it was officially added to the British list in 1971 and now numbers around 7,000 birds.

HABITAT Well-vegetated lakes and rivers set in woodland – it nests in holes in trees.

VOICE Rather silent.

CONFUSION SPECIES Drake unique, duck easily confused with Wood Duck *A. sponsa*, a native of America that occasionally escapes, but Mandarin has narrower white line across head, which is paler, more boldly spotted flanks, and pale rather than dark 'nail' at tip of bill.

male

eclipse male

female

The male is handsome in bottle-green, chestnut and white, but the female is very like a female Mallard and best picked out by the shovel-like bill.

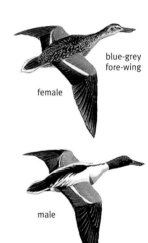

female

blue-grey fore-wing

male

DESCRIPTION Slightly smaller than Mallard, bill expanded and flattened towards tip – an adaptation for filter-feeding; legs orange. Male unmistakable; plumage of female similar to Mallard although head plainer; in flight dark belly contrasts with white underwing, forewing dull greyish and speculum lacks white trailing edge. Eclipse male similar to female but darker and more rufous on head, flanks and belly, with pale eye and pale blue forewing. Juvenile slightly darker than female. Generally found in pairs and small parties.

POPULATION Scarce breeder, scattered throughout British Isles but mostly in E; most breeding birds move to SW Europe in winter, to be replaced by significantly greater numbers of birds from Europe and Siberia.

HABITAT Breeds in rough grassland around shallow open water and marshy pools; winters on well-vegetated lakes, reservoirs and gravel pits, less commonly around estuaries.

VOICE A clipping *chuk-chuk-chuk-chuk*.

CONFUSION SPECIES Breeding male distinctive, otherwise very like Mallard, but bill shape always diagnostic.

GADWALL · 51 CM · *Anas strepera*

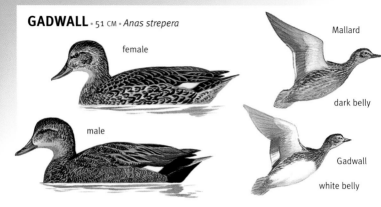

female

male

Mallard

dark belly

Gadwall

white belly

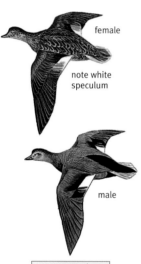

female

note white speculum

male

Widespread in small numbers, the male is subtly attractive, but the female resembles a Mallard and is easily overlooked.

DESCRIPTION Slightly smaller and slimmer than Mallard. In all plumages has small square of white on inner wing, visible on water and in flight. Grey body and black rear end of winter male distinctive. Female very like female Mallard but note white in wing, white belly and dark bill with neat orange stripe along cutting edges. Eclipse male similar to female but retains grey wings. Juveniles hard to identify, as adult female but white in wing much reduced, especially in females, and belly mottled. Usually seen in pairs or small parties, only occasionally in large flocks.

POPULATION Widespread but uncommon breeding species, mostly originating from birds introduced to Norfolk in the 19th century, but natural populations probably occur in N Scotland. Numbers and range have steadily expanded in recent decades. Some British breeders winter in Europe, to be replaced by immigrants from Iceland, Scandinavia and the Low Countries.

HABITAT Breeds and winters around shallow, well-vegetated freshwaters, uncommonly also on estuaries.

VOICE Pure, thin whistled *pee* and more quacking *ak, ak*.

CONFUSION SPECIES Female Mallard.

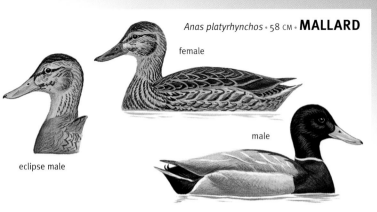

Anas platyrhynchos • 58 CM • **MALLARD**

female

male

eclipse male

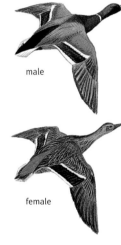

male

female

The wild duck, common on almost any waterbody and often tame, but beware lookalike domestic stock, especially on village ponds and park lakes.

DESCRIPTION Winter male distinctive. Female variegated brown, with blue speculum broadly bordered white and often visible on closed wing. Legs orange and bill dark, irregularly grading to orange at sides. Eclipse male as female but bill yellow and breast rufous-brown; juvenile as female. In flight note white underwing. Often rather tame, but unusually dark or pale plumage, a pale breast, etc., indicates the presence of genes from domestic breeds.

POPULATION Abundant resident, numbers are supplemented in winter by visitors from Iceland and Scandinavia.

HABITAT Almost any type of freshwater habitat, even very small ponds and ditches, also estuaries and sheltered coastal waters.

VOICE Male gives a soft, nasal *kreep*, female gives the classic *quak*, often in a 'laughing' series: *quak, quak, quak-quak-quak-quak*.

CONFUSION SPECIES Female and immature Mallards are easily confused with Gadwall, less so with Shoveler, Wigeon and Pintail, and even the rather smaller Teal. Look for head and bill shape, leg and bill colour, wing pattern and overall tone of plumage.

PINTAIL · 51–66 cm · *Anas acuta*

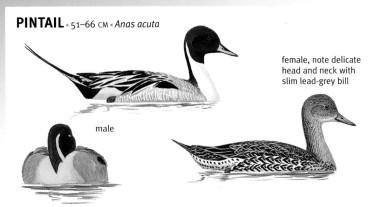

female, note delicate head and neck with slim lead-grey bill

male

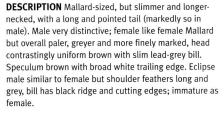

female

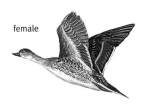

male

A supremely elegant duck, odd birds can be seen almost anywhere in winter but large flocks are very local in occurrence.

DESCRIPTION Mallard-sized, but slimmer and longer-necked, with a long and pointed tail (markedly so in male). Male very distinctive; female like female Mallard but overall paler, greyer and more finely marked, head contrastingly uniform brown with slim lead-grey bill. Speculum brown with broad white trailing edge. Eclipse male similar to female but shoulder feathers long and grey, bill has black ridge and cutting edges; immature as female.

POPULATION Very rare, local and sporadic breeder, rather more numerous in winter, when several tens of thousands arrive from the Continent.

HABITAT In winter found on lakes, reservoirs and gravel pits, but the vast majority winter on estuaries, especially the Mersey and Dee in NW England. Breeds on shallow pools in grasslands.

VOICE Call a mellow *proop*.

CONFUSION SPECIES Female similar to female Gadwall and Mallard and, to a lesser extent, female Wigeon and Shoveler, but note grey bill and legs.

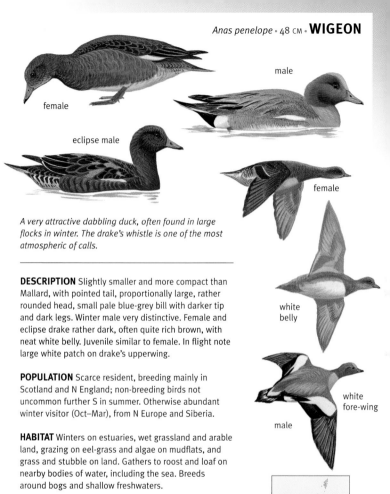

Anas penelope • 48 CM • **WIGEON**

female

male

eclipse male

female

white belly

white fore-wing

male

A very attractive dabbling duck, often found in large flocks in winter. The drake's whistle is one of the most atmospheric of calls.

DESCRIPTION Slightly smaller and more compact than Mallard, with pointed tail, proportionally large, rather rounded head, small pale blue-grey bill with darker tip and dark legs. Winter male very distinctive. Female and eclipse drake rather dark, often quite rich brown, with neat white belly. Juvenile similar to female. In flight note large white patch on drake's upperwing.

POPULATION Scarce resident, breeding mainly in Scotland and N England; non-breeding birds not uncommon further S in summer. Otherwise abundant winter visitor (Oct–Mar), from N Europe and Siberia.

HABITAT Winters on estuaries, wet grassland and arable land, grazing on eel-grass and algae on mudflats, and grass and stubble on land. Gathers to roost and loaf on nearby bodies of water, including the sea. Breeds around bogs and shallow freshwaters.

VOICE Drake gives a loud and shrill, whistled *pee-oow* or *p'pee-oow*.

CONFUSION SPECIES Female, eclipse male and immature distinguished from other brownish ducks by head shape (especially steep forehead, high, rounded crown and small bill), rather uniform, often rich brown coloration and white belly.

GARGANEY · 39 CM · *Anas querquedula*

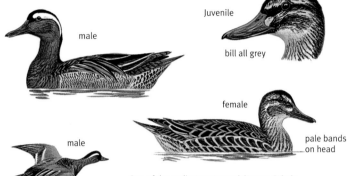

male

Juvenile

bill all grey

male

female

pale bands on head

One of the earliest summer visitors to Britain, arriving in March, this secretive duck is usually seen singly or in pairs.

juvenile male (left) female (right)

DESCRIPTION Slightly larger than Teal. Drake distinctive, with broad white line from just in front of eye backwards across brown head and, in flight, pale blue-grey fore-wing and marked contrast between the dark breast and white belly. Females and non-breeding birds very like Teal but with longer, all-grey bill, bolder head pattern (with dark line from bill back through eye, bordered above and below by paler lines, and diffused pale spot at base of bill), and also often whitish throat. In flight no green in speculum.

POPULATION Typically fewer than 100 pairs breed. Rather scarce on passage in spring and Aug–Sep; winters in central Africa.

HABITAT Shallow, well-vegetated wetlands, on passage also lakes, reservoirs and gravel pits. Always on fresh water.

VOICE Drakes have a distinctive display call, a dry, wooden rattle, *k-r-r-r-r-r-r-eet, k-r-r-r-r-r-r-eet*....

CONFUSION SPECIES Females and autumn birds very easily confused with Teal.

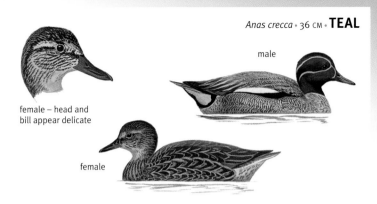

male

female – head and
bill appear delicate

female

Teal are shy and nervous (even their calls sound nervous). If surprised flocks will explode upwards off the water in a 'vertical take-off'.

male

DESCRIPTION Smallest dabbling duck, speculum dark green bordered white, broadly so on leading edge. Male distinctive; female very like a small female Mallard, but note delicate bill, mostly dark with a little orange-yellow at base, dark legs, green speculum (if visible) and short horizontal white line at base of tail. Eclipse male and juvenile as female. Very sociable; in flight forms tight flocks.

female

POPULATION A declining breeder in N England, Scotland and N Ireland, rare further south. Numbers augmented in winter by birds from N Europe and Siberia, and then abundant in much of Britain and Ireland.

HABITAT Breeds on moorland and bog pools in N and W, also well-vegetated waters in lowlands. In winter found on estuaries and sheltered coastal waters, lakes, reservoirs, gravel pits, flooded grassland and even quite small ponds.

VOICE Call a nervous, slightly tremulous, whistled *cree*.

CONFUSION SPECIES Small size distinctive, apart from Garganey, which is almost as small, but female Teal has a rather plainer head and duller wing pattern.

POCHARD · 46 CM · *Aythya ferina*

eclipse male

male

head shape
always
distinctive

female

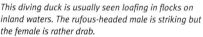

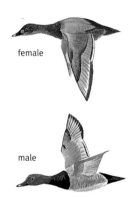

female

male

This diving duck is usually seen loafing in flocks on inland waters. The rufous-headed male is striking but the female is rather drab.

DESCRIPTION Slightly larger than Tufted Duck, head shape distinctive, with high, peaked crown and sloping forehead that grades into the long, concave, deep-based bill. In all plumages shows broad greyish wing-bar. Female nondescript greyish-brown with darker and browner head, breast and rear end and vague pattern on face – pale spot at base of bill, eye-ring and streak back from eye. Juvenile similar but even duller. Dives for food as well as up-ending and dabbling.

POPULATION Scarce breeding resident, mostly in England and E Scotland. Much commoner in winter, when large influxes from central Europe and Russia occur. Often found in sizeable flocks.

HABITAT Breeds around well-vegetated freshwaters; winters on lakes, reservoirs and gravel pits.

VOICE Generally silent.

CONFUSION SPECIES Head shape and dull grey wing-bars are good pointers in all plumages. Head of male completely reddish, unlike drake Teal and Wigeon.

male Tufted Duck

Aythya fuligula • 44 CM • **TUFTED DUCK**

Aythya marila • 47 CM • **SCAUP**

female Tufted Duck

male Scaup

female Scaup summer

Tufted Duck is the commonest diving duck; the tuft that gives it its name is the small crest on the back of the head. The closely-related Scaup is an uncommon winter visitor.

female Tufted Duck

male Tufted Duck

DESCRIPTION Breeding males distinctive. Females and eclipse males much more similar: Tufted Duck always shows at least a hint of a tuft on the head and a grey bill with a 'dipped in ink' black tip. Scaup has much less black – just a black 'nail' – and a broad white blaze around the base of bill (female Tufted may show white here, but it is usually much less obvious). Scaup is also slightly bigger and broader in the beam, with a smoothly rounded crown.

POPULATION Tufted Duck is a common resident, with much larger numbers arriving from Europe to winter. Scaup is rather scarcer: fewer than 10,000 winter and just 1–2 pairs occasionally breed.

HABITAT Tufted Duck is found on lakes, reservoirs, gravel pits and ponds, with small numbers also on sheltered estuaries, and breeds around well-vegetated freshwater. Scaup favours sheltered coastal waters, such as Dee estuary, Solway Firth, Firth of Forth and Moray Firth, but is also found inland in very small numbers. Mostly found on salt water.

VOICE In display male Tufted Duck gives a series of rapid bubbling notes, female has growling call.

CONFUSION SPECIES See Pochard.

Tufted Duck

Scaup

COMMON SCOTER · 49 CM · *Melanitta nigra*
VELVET SCOTER · 55 CM · *Melanitta fusca*

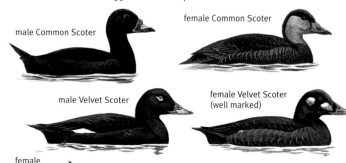

male Common Scoter

female Common Scoter

male Velvet Scoter

female Velvet Scoter (well marked)

female Common Scoter

female Velvet Scoter

Common Scoter

Velvet Scoter

Common Scoter is the most widespread sea duck, found off almost all coasts and at any time of year, and it is always worth checking for the much scarcer Velvet Scoter.

DESCRIPTION Scoters form large flocks, often well offshore and visible only as black dots on the sea or flying low in long, straggly lines. They often swim with tail cocked and dive for food with a small leap. Drakes are black: male Common Scoter has a black bill with swollen base and yellow central ridge (hard to see); male Velvet has extensive orange sides to bill and tiny white 'tick-mark' below eye. Females and immatures are sooty-brown: female Common has paler cheeks, female Velvet smudgy pale spots at base of bill and on ear-coverts. At all times Velvet Scoter is best identified by its white inner flight feathers.

POPULATION Common Scoter is a fairly common winter visitor, passage migrant and nonbreeding summer visitor from N Europe, to the coast of Wales, NE Scotland and N Norfolk. Scarce inland on passage and in winter, and a rare breeder, with 50 pairs each in N Scotland and Ireland; it is on the conservation 'Red List'. Around 3,000 Velvet Scoters winter, mostly off NE coasts; very rare inland.

HABITAT Both species winter in sheltered coastal waters. Common Scoter breeds around lakes and bogs.

VOICE Generally silent.

CONFUSION SPECIES See Eider and Scaup.

female

Somateria mollissima • 61 CM • **EIDER**

male

eclipse male

1st-winter male

This northern duck breeds in N Ireland, Scotland and N England. 'Eider down' is the breast feathers used by female to line the nest.

DESCRIPTION Large, heavy duck with thick neck and wedge-shaped head and bill. Black and white male distinctive, note pink flush on breast and green sides to head, but in eclipse plumage much drabber, mostly sooty-black. First-year male similarly dull, but has white breast. Female warm brown with black vermiculations, juvenile more uniformly grey-brown. Dives for food, primarily mussels.

POPULATION Common breeding resident (south to Northumberland and Cumbria), numbers have slowly increased in recent decades. Most do not wander far from breeding areas in winter, but small numbers arrive from Scandinavia to winter along E and S coasts of England.

HABITAT Sheltered coastal bays and estuaries, spending much time loafing on beaches, rocky reefs and sandbars. Strictly marine and very rare inland.

VOICE Courting male has wonderful, rolling and rather pigeon-like *aa-ooow, aa-ooow…* given as it throws its head back.

CONFUSION SPECIES Female could be confused with female Mallard, which often rests on the sea, but is much more heavily built.

female

male

LONG-TAILED DUCK · 40–47 CM · *Clangula hyemalis*

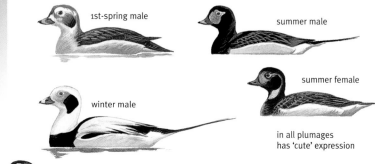

1st-spring male

summer male

summer female

in all plumages
has 'cute' expression

winter male

winter female

A small sea duck, reasonably common off Scottish coasts in winter but rather scarce further south and rare inland.

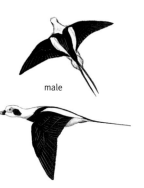

male

DESCRIPTION Size as Tufted Duck, rather 'neckless' with a rounded head and short, stubby bill; male has extremely long, thin central tail feathers (up to 13 cm). Complex pattern of moults produces at least three plumages each year, combinations of brown, black and white, but always has 'cute' expression and, in flight, all-dark wings. Winter male largely white, with black breast and cheek patch, female dusky, with whitish flanks, neck and face and dark cap and cheek patch. Dives for food, and usually seen in small, scattered flocks.

POPULATION Winter visitor from the Arctic, numbers peaking late Dec–Feb, but the majority are usually far offshore and out of sight.

HABITAT Winters at sea, in sheltered bays and estuaries, favouring shallow water over sandy and muddy substrates. Inland birds favour lakes, reservoirs and gravel pits.

VOICE Call *ah-oowa-lip*, vaguely goose-like, given in chorus by displaying males in late winter.

CONFUSION SPECIES Despite complex plumages, size and shape are distinctive and separation from other ducks usually straightforward if views good. May recall an auk, even a winter-plumaged Puffin.

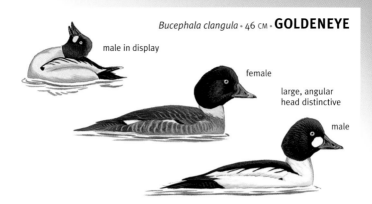

Bucephala clangula • 46 CM • **GOLDENEYE**

male in display

female

large, angular
head distinctive

male

This diving duck is a boreal species that has colonised the Scottish Highlands following the provision of special nest boxes.

DESCRIPTION Size as Tufted Duck but body tapered at rear with a large, angular head and small triangular bill; in flight much white on inner wing. Male distinctive. Female greyish with square white patch towards rear of body, whitish collar and rich brown head; eclipse male and juvenile similar to female, but juvenile has dark rather than golden eye and lacks pale collar. Often seen in pairs and small parties and generally very active, spending much of time diving for food.

POPULATION First bred in Scotland in 1970 and has increased to over 100 pairs, mostly in Badenoch and Strathspey. Otherwise fairly common winter visitor from N Europe, although sparse in SW England and Ireland.

HABITAT Breeds near forested lakes and rivers, using tree holes or, much more commonly, nest boxes. Winters on lakes, reservoirs and gravel pits, also on estuaries and the sea.

VOICE Displaying male throws head back and gives a rapid quack. In flight the wings produce a loud musical whistle, especially in males.

CONFUSION SPECIES Immatures can be very drab and are best identified by head shape and white patch on rear body.

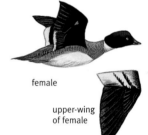

female

upper-wing
of female

male

SMEW · 41CM · *Mergus albellus*

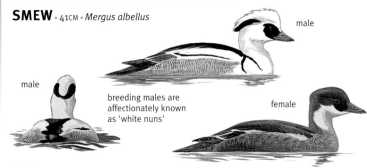

male

male

breeding males are
affectionately known
as 'white nuns'

female

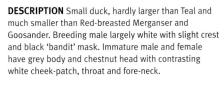

females (shown) and immature
males are known as 'redheads'

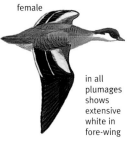

female

in all
plumages
shows
extensive
white in
fore-wing

male

*A scarce visitor, this dainty sawbill is very much a
'blue riband' find for birdwatchers out on a cold
winter's day.*

DESCRIPTION Small duck, hardly larger than Teal and
much smaller than Red-breasted Merganser and
Goosander. Breeding male largely white with slight crest
and black 'bandit' mask. Immature male and female
have grey body and chestnut head with contrasting
white cheek-patch, throat and fore-neck.

POPULATION Winter visitor late Nov–Mar from northern
Scandinavia and Siberia, with most in SE England.
Usually under 100 birds, but sometimes more if weather
in Europe particularly cold.

HABITAT Lakes, reservoir and gravel pits.

VOICE Usually silent.

CONFUSION SPECIES Adult male distinctive, but
female and immature male easily confused with similarly
white-cheeked Ruddy Duck, *Oxyura jamaicensis*
(accidentally introduced to Britain in 1950s). Ruddy duck
has similar white cheek patches, contrasting with a
chestnut (breeding male) or sooty-brown (other
plumages) body. Note, however, its long tail, often held
cocked, lack of white in the wing and, in all plumages,
blackish head.

male

female has rugged
double crest

female

This northern duck is a 'sawbill', the serrated edge to its bill being an adaptation to catch and hold fish, and its fish-eating habits make it unpopular in some quarters.

eclipse male

DESCRIPTION Mallard-sized, with long and slender red bill, red eye and short, ragged double crest. In flight white patch on inner wing is crossed by black lines. Male distinctive; lower breast black, blotched white at sides, upper breast and neck cinnamon, spotted black, separated from bottle-green head by white collar. Female ash-grey with rufous upper neck and head. Juvenile and eclipse male similar to female, but latter has extensive white in wing. Often found in small flocks. Dives for food.

female

POPULATION Fairly common breeder in N and W, but numbers breeding inland have fallen in recent years; most winter on coasts near breeding areas. Further S occurs in small numbers around all coasts (Jul–Mar), these birds probably of European origin.

male

HABITAT Breeds on sheltered coasts, sea lochs and estuaries, rather less commonly along rivers. Nests on the ground. Winters in similar habitats, but rare inland except during occasional hard weather influxes.

VOICE Usually silent.

CONFUSION SPECIES Female very like Goosander, but rufous of head not sharply divided from grey of neck and lacks well-defined white chin, head also paler rufous.

GOOSANDER · 62 CM · *Mergus merganser*

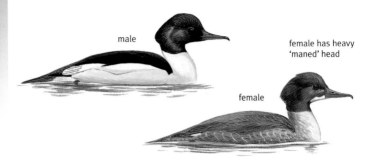

male

female has heavy 'maned' head

female

juvenile

female

male

Males of this 'sawbill' are distinctive, but separation of females from Red-breasted Merganser requires care, although it is much the commoner on freshwater.

DESCRIPTION Slightly larger than Mallard with long, hook-tipped red bill, dark eye and short, full crest giving a big-headed appearance. In flight shows white on inner wing, extensive in male. Male distinctive. Female greyish with reddish-brown head, juvenile similar but duller, brown of head not so well-defined, with narrow white line from eye to bill and yellowish eye. Eclipse male as female but retains extensive white in wing. Dives for food.

POPULATION Fairly common breeder. First recorded nesting in Scotland in 1871 and has expanded S into Wales, the Peak District and SW England in recent decades. Fairly common in winter throughout Britain, birds in S and E being mostly immigrants from Europe.

HABITAT Breeds along upland rivers and also around forested lakes, nesting in a tree hole, crevice or nest box. Winters on freshwater, on larger lakes, reservoirs and gravel pits, sometimes also on very sheltered coasts.

VOICE Harsh *krak-krak-krak…* in display.

CONFUSION SPECIES Female like Red-breasted Merganser, but note neater and darker brown head clearly separated from grey neck and neat white chin, and cleaner, purer grey body.

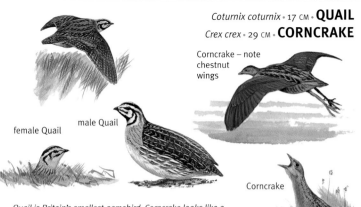

Coturnix coturnix • 17 CM • **QUAIL**
Crex crex • 29 CM • **CORNCRAKE**

Corncrake – note chestnut wings

female Quail

male Quail

Corncrake

Quail is Britain's smallest gamebird. Corncrake looks like a gamebird but is in fact closely related to Water Rail. Both are small birds that like dense cover and are hard to see.

DESCRIPTION Quail is tiny, with striped head and, in males, black throat. Most views are in flight, which is fast, low and level. Corncrake is very secretive, but calling males are sometimes bolder. Flight fluttering, with dangling legs and prominent chestnut wings.

POPULATION Numbers of Quail are variable: few in poor years, mostly in S England, but up to 1,500 calling birds in good years, when much more widespread. Summer visitor from Africa, May onwards. Corncrake numbers dropped to 489 singing males in 1993 but, following habitat management, now around 1,200, with an on-going reintroduction programme in Cambridgeshire. Visitor from tropical Africa, mid Apr–Sep.

HABITAT Both keep to dense cover. Quail favours grassland and cereal fields. Corncrake is largely confined to damp meadows in the Hebrides and is on the conservation 'Red List'.

VOICE Quail call is a liquid *quip-ip-ip* ('wet my lips'), often given at night. Corncrake's is a rasping *crex-crex-crex...* (hence scientific name), like fingernails run across a comb, and it also often calls at night.

CONFUSION SPECIES No gamebird is as small as Quail. In its dry habitat, Corncrake is unmistakable.

Quail

Corncrake

49

RED GROUSE · 40 CM · *Lagopus lagopus*

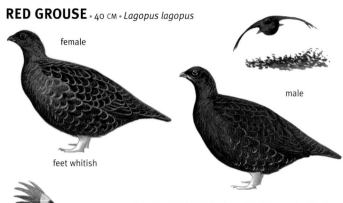

female

male

feet whitish

Red Grouse can easily hide in the heather

A much-prized gamebird, confined to moorland in the N and W of Britain but in long-term decline and now rare in Ireland.

DESCRIPTION Slightly larger than a partridge. Rusty-brown all over, finely barred and vermiculated, with whitish feet and black tail. In flight shows white under-wing. Males on average darker, with a tiny white streak at base of bill and, in spring, bright red wattle over eye. Female has more conspicuous yellowish feather fringes, forming pale scallops, and the wattle much reduced or absent. Continental populations, which become all-white in winter, are known as 'Willow Grouse'.

POPULATION Fairly common resident in upland areas. Many moors are managed to encourage grouse by burning the heather in rotation and controlling predators, but, unlike Pheasant, captive-bred birds are not released and most grouse moors undergo natural 4–8 year population cycles.

HABITAT Open heather moorland, occasionally also farmland in hard weather.

VOICE A loud *krrrr-k-k-k-k*, *go-back*, *go-back*....

CONFUSION SPECIES Ptarmigan always shows white on wings and belly and is all-white in winter; female Black Grouse has narrow whitish wing-bar, pale vent and finely barred tail.

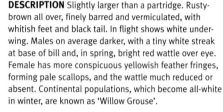

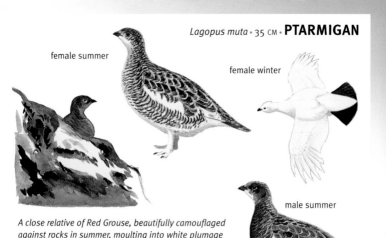

Lagopus muta • 35 CM • **PTARMIGAN**

female summer

female winter

male summer

A close relative of Red Grouse, beautifully camouflaged against rocks in summer, moulting into white plumage for snowy winters.

DESCRIPTION In summer most of plumage closely vermiculated, overall tone greyish in male and yellowish-grey in female, with belly and wings white – the latter very conspicuous in flight. Males have red wattle above eye, most obvious in spring. In winter becomes white, apart from black tail and, in male, black lores, while in spring and autumn shows pied, intermediate plumage. Often very tame and can be extremely hard to spot when sitting still.

POPULATION Fairly common in right habitat, with UK population around 10,000 pairs.

HABITAT The tundra-like tops of Scottish mountains, with plenty of boulders and short vegetation dominated by lichens, sedges and dwarf shrubs. Confined to high peaks in S of range, but may be found near sea level in extreme NW Scotland.

male winter

VOICE A dry, croaking rattle (like stick pulled rapidly across slats of picket fence).

CONFUSION SPECIES Red Grouse is slightly larger, rusty-brown rather than greyish, and never shows extensive white in the wings or a white belly.

CAPERCAILLIE · 60–87 CM · *Tetrao urogallus*
BLACK GROUSE · 40–55 CM · *Tetrao tetrix*

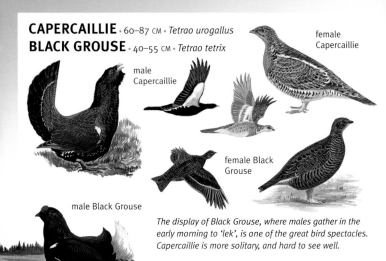

male Capercaillie

female Capercaillie

female Black Grouse

male Black Grouse

The display of Black Grouse, where males gather in the early morning to 'lek', is one of the great bird spectacles. Capercaillie is more solitary, and hard to see well.

DESCRIPTION Male Capercaillie is Turkey-sized with long fan-shaped tail, female smaller. Male Black Grouse is black with extensive white in under-tail and in wing, red wattle and lyre-shaped tail. Female Black Grouse is closely barred and appears grey-brown at rest; it lacks female Capercaillie's rusty throat and breast, but shows a white vent and, in flight, narrow white wingbars.

POPULATION Capercaillie was hunted to extinction in Scotland by 1770s. It was re-introduced in 1837 and thrived for a while, but has recently declined rapidly to around 1,200 birds and is in real danger of a second extinction. Black Grouse has also declined recently and recent counts give a UK population of 5,000 lekking males. Both species are on the conservation 'Red List'.

HABITAT Capercaillie is found in pine woodland and pine plantations, with Black Grouse favouring a mosaic of moorland, rushy pastures and woodland.

VOICE Capercaillie's song is quiet throaty clicks followed by a noise like a cork being popped. Lekking Black Grouse give a far-carrying, eerie, bubbling moan, *oodle-u, oodle-u, woooo, woooo, oodle-u...*, intermixed with harsh hisses.

Capercaillie

Black Grouse

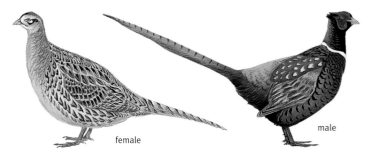

female

male

A common and striking gamebird, the male's crowing is evocative of the British countryside, but Pheasants are not native, being ancient introductions from Asia.

DESCRIPTION Male has glossy green head with extensive bare red skin. Otherwise rather variable, many have a white collar and greyish rump, some are rather dark over whole head and breast, this variation reflecting the varied origins of captive-bred stock. Female uniformly brown with dark spots and chevrons, juvenile similar but smaller and shorter-tailed.

males explode upwards when flushed

POPULATION Common to abundant resident. Probably introduced to England by the Normans in the 11th or 12th century, possibly even earlier, by the Romans. Introduced to Ireland in the 16th century. The natural range extends from the Black Sea to China. Large numbers of captive-bred birds are released every autumn to boost numbers for the winter shoots.

female

HABITAT A mixture of farmland and woodland, also reedbeds; will visit nearby gardens.

VOICE Call a single crowing *kook* or *kru-kook*, accelerating into a cackle in flight.

CONFUSION SPECIES Pheasant chicks vaguely recall Grey Partridge but have a longer, more pointed all-brown tail.

GREY PARTRIDGE • 30 CM • *Perdix perdix*

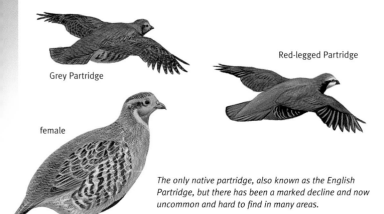

Grey Partridge

Red-legged Partridge

female

The only native partridge, also known as the English Partridge, but there has been a marked decline and now uncommon and hard to find in many areas.

male

DESCRIPTION Adult has upperparts finely barred and streaked; face ochre and underparts grey with a dark horseshoe-shaped patch on breast and rusty bars on flanks. Bill and legs greyish. Sexes similar, although female a little duller. Juvenile very different, overall streaky brown. Usually seen in coveys. Shy, it tends to crouch or run away or, if surprised, the whole flock explodes into the air and flies off with whirring wings and long glides.

POPULATION Once abundant, it has declined severely in recent decades, especially in the N and W, to the extent that is now uncommon and on the conservation 'Red List' in Britain; almost extinct in Ireland. Resident, only ever moving short distances.

HABITAT Farmland, especially with hedgerows and other shelter.

VOICE A single, prolonged grating *kree'eet*, often given in the early morning and evening.

CONFUSION SPECIES Juveniles can be mistaken for a Pheasant chick or even a Quail, but usually stay with adults. See also Red-legged Partridge.

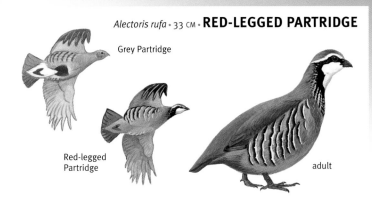

Grey Partridge

Red-legged Partridge

adult

Alectoris rufa • 33 CM • **RED-LEGGED PARTRIDGE**

Introduced in the 18th century, in most of Britain the 'French Partridge' is now much commoner than its native cousin, the Grey Partridge.

DESCRIPTION Sexes similar. Juveniles much duller. Usually found in pairs or small coveys, running off when disturbed or, when pressed harder, exploding into the air and flying away.

POPULATION Introduced from SW Europe and now a common resident, but has declined in recent decades, in part due to the release prior to 1992 of large numbers of hybrid Red-legged Partridge × Chukar *A. chukar* (a related species from eastern Europe), which do not breed very successfully in the wild.

HABITAT Farmland, where favours arable crops, especially sugar beet. Prefers the presence of some cover, such as hedgerows or scrub, and may also be found on heathland or even in open woodland.

VOICE Calls frequently, a rhythmic chicken-like *kuk'kuk-kacha*.

CONFUSION SPECIES Easily separated from Grey Partridge by black and white face pattern, uniform upperparts, boldly barred flanks and red bill and legs; in flight note also grey flight feathers and plain grey rump and centre to tail (sides of tail rufous in both species).

Chukar

Red-legged Partridge

GREAT CRESTED GREBE · 49 CM · *Podiceps cristatus*

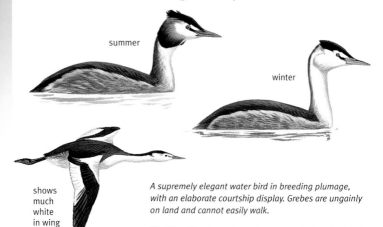

summer

winter

shows
much
white
in wing

winter

juvenile

A supremely elegant water bird in breeding plumage, with an elaborate courtship display. Grebes are ungainly on land and cannot easily walk.

DESCRIPTION Largest grebe, near Mallard in size. In all plumages note rather long, thin neck, pinkish bill and, in flight, striking white wing-patches. In winter upperparts grey-brown, underparts and face white, including area above eye. Sexes similar. Juvenile has dark stripes on neck. Dives underwater in search of fish.

POPULATION Fairly common resident, with some immigration from Europe in winter. In the 19th century reduced to just 42 pairs in England by trade in the head plumes, known as 'grebe fur', but numbers have recovered to around 5,000 pairs.

HABITAT Breeds on medium to large, shallow and well-vegetated freshwater lakes, reservoirs and gravel pits. Nest a floating mat of vegetation. Adults may carry their striped chicks piggyback as they swim. Winters in similar habitats, as well as sheltered inshore waters.

VOICE Guttural croaks and chatters.

CONFUSION SPECIES Much larger than Little Grebe. In winter close to Red-necked Grebe *P. grisegena*, a scarce migrant and winter visitor, but note latter's yellow-based bill and dusky cheeks and throat. Separated from winter Red-throated Diver by silhouette, especially more angular head, whiter face and pinkish bill.

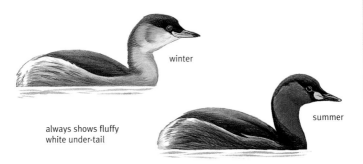

winter

always shows fluffy
white under-tail

summer

*Affectionately known as 'Dabchick', this is the smallest
grebe, breeding on well-vegetated freshwaters and
drawing attention with its whinnying call.*

DESCRIPTION A little smaller than Moorhen, rotund and
appearing very buoyant. Note fluffy white undertail. In
breeding plumage has sooty upperparts, head and neck,
with chestnut foreneck and contrasting yellow spot at
bill-base. Sexes similar. In winter dull brown above and
buffy below, with pale bill. Juvenile as adult winter but
has striped head and neck. Dives frequently and often
skulks in waterside vegetation.

scutters across water

POPULATION Fairly common resident, numbers are
supplemented in winter by Continental immigrants.

HABITAT When breeding prefers smaller waters with
plenty of cover, sometimes even just a ditch. In winter
more widespread and also found on larger, more open
waterbodies and sheltered inshore waters; may form
quite large flocks.

winter

VOICE Loud, often prolonged, whinnying trill that rises
and falls in volume and sometimes also in pitch.

CONFUSION SPECIES Small size distinctive. Separated
in non-breeding plumage from slightly larger Black-
necked and Slavonian Grebes *Podiceps nigricollis* and
P. auritus (both rare breeders and scarce migrants/winter
visitors) by brown and buff, rather than dark grey and
white, coloration.

RED-THROATED DIVER · 53–69 CM · *Gavia stellata*

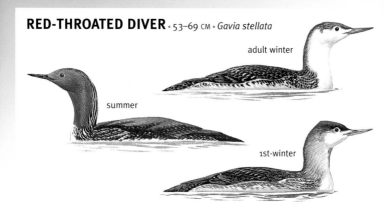

adult winter

summer

1st-winter

often flies high

bill slender
and often
held tilted
slightly upwards

Divers are named after their habit of diving to feed and can spend up to 60 seconds submerged. Their alternative American name of 'loon' refers to their eerie songs.

DESCRIPTION Mallard-sized, legs set far back on body, making efficient paddles but of little use on land. In breeding plumage has distinctive but surprisingly hard to see brick-red foreneck. In winter grey above and white below, with rather white face. Sexes similar; immature similar to adult winter.

POPULATION Scarce breeding bird in NW Scotland with a few in NW Ireland. Fairly common in winter, especially on E coast, numbers are supplemented by birds from Iceland and Scandinavia.

HABITAT Breeds on small moorland lochans, flying to larger lochs or the sea to feed. In winter strictly marine, favouring shallow, sheltered inshore waters, often forming loose flocks; very rare inland.

VOICE Flight call on breeding grounds a cackling *cak cak*. Song a goose-like honking and tremulous wailing. Silent in winter.

CONFUSION SPECIES Careful study of overall silhouette and size and shape of bill should separate it from Cormorant, Shag, sea ducks and auks. Harder to separate from scarce Great Northern and Black-throated Divers *G. immer* and *G. arctica*, but note slender bill, often held pointing slightly upwards.

sits high on water

distinctive stiff-winged flight

Closely related to albatrosses, this is one of a group of seabirds know as 'tubenoses'; their tube-shaped nostrils are an adaptation to excrete excess salt.

DESCRIPTION Size as Common Gull, plumage gull-like but rump and tail grey and no black on wing-tips, rather a pale flash at base of primaries. Bill heavy, ochre towards tip. Sexes and ages similar. Usually seen flying to and fro around breeding colonies. In strong winds glides effortlessly for long distances, keeping low over waves. When settled, sits high on water.

POPULATION Numbers and range expanded dramatically in the late 19th and 20th centuries and now common, with over 500,000 pairs in the British Isles. Birds remain near breeding colonies throughout the year, but also disperse more widely and can be seen off all coasts.

HABITAT Breeds on sea cliffs, sometimes also buildings close to the sea and even quarries and crags up to 20km inland. Otherwise, spends its entire life at sea, the rare inland records away from breeding colonies mostly involve storm-blown birds.

VOICE A croaking *gak-gak-gak* at nest.

CONFUSION SPECIES Best separated from gulls by flight action, a mixture of glides in which the wings are held rigidly straight outwards, and flapping with a rapid and shallow, stiff-winged motion.

MANX SHEARWATER · 34 CM · *Puffinus puffinus*

in distant views
flashes black then
white as it banks

*After an absence of many years, this seabird has recently
returned to the island that gave it its name, breeding
again on the Calf of Man.*

DESCRIPTION Size near Black-headed Gull. Sooty-black
above, white below. Totally pelagic away from breeding
colonies, keeps well out to sea and unlikely to be seen
unless 'seawatching'. In calm weather flies low over
water, bouts of flapping on stiff, straight wings
interspersed with long glides on rigid, slightly down-
curved wings. In high winds mostly glides, showing
alternately black and white as it changes angle.

POPULATION With 300,000 breeding pairs, the British
Isles support over 80% of the world population.
Abundant summer visitor (late Mar–Sept). Winters in
South Atlantic.

HABITAT Breeds in huge colonies in burrows on around
40 offshore islands. Birds gather nearby towards dusk,
flying around and settling on sea, but only come ashore
once darkness has fallen. On passage moves south,
often in large numbers on W coasts of Britain and
Ireland, but elsewhere usually well out to sea unless
pushed inshore by strong winds.

VOICE In breeding colonies at night gives bizarre hoarse
wailing.

CONFUSION SPECIES Black and white plumage, flight
action and habitat unique.

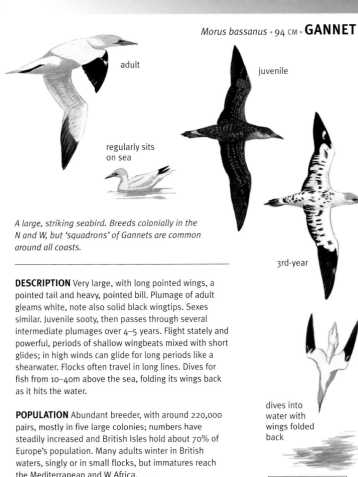

Morus bassanus • 94 CM • **GANNET**

adult

juvenile

regularly sits
on sea

3rd-year

*A large, striking seabird. Breeds colonially in the
N and W, but 'squadrons' of Gannets are common
around all coasts.*

DESCRIPTION Very large, with long pointed wings, a
pointed tail and heavy, pointed bill. Plumage of adult
gleams white, note also solid black wingtips. Sexes
similar. Juvenile sooty, then passes through several
intermediate plumages over 4–5 years. Flight stately and
powerful, periods of shallow wingbeats mixed with short
glides; in high winds can glide for long periods like a
shearwater. Flocks often travel in long lines. Dives for
fish from 10–40m above the sea, folding its wings back
as it hits the water.

POPULATION Abundant breeder, with around 220,000
pairs, mostly in five large colonies; numbers have
steadily increased and British Isles hold about 70% of
Europe's population. Many adults winter in British
waters, singly or in small flocks, but immatures reach
the Mediterranean and W Africa.

HABITAT Breeds on rocky cliffs and slopes, mostly on
offshore islands, northwards from Bempton Cliffs
(Yorkshire) in E and Grassholm (Pembrokeshire) in W.
Otherwise, spends its entire life at sea. Very rare inland.

VOICE A hoarse *aar-r-r* at breeding colonies.

CONFUSION SPECIES Shape and flight very distinctive,
as is very white plumage of adult.

dives into
water with
wings folded
back

SHAG · 73 CM · *Phalacrocorax aristotelis*

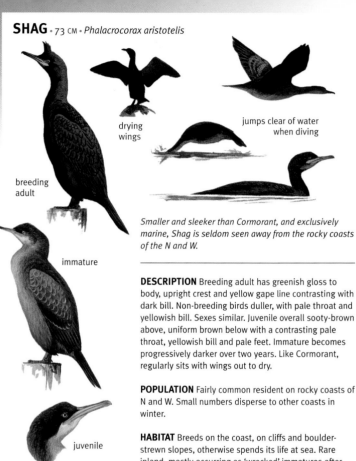

drying wings

jumps clear of water when diving

breeding adult

immature

juvenile

Smaller and sleeker than Cormorant, and exclusively marine, Shag is seldom seen away from the rocky coasts of the N and W.

DESCRIPTION Breeding adult has greenish gloss to body, upright crest and yellow gape line contrasting with dark bill. Non-breeding birds duller, with pale throat and yellowish bill. Sexes similar. Juvenile overall sooty-brown above, uniform brown below with a contrasting pale throat, yellowish bill and pale feet. Immature becomes progressively darker over two years. Like Cormorant, regularly sits with wings out to dry.

POPULATION Fairly common resident on rocky coasts of N and W. Small numbers disperse to other coasts in winter.

HABITAT Breeds on the coast, on cliffs and boulder-strewn slopes, otherwise spends its life at sea. Rare inland, mostly occurring as 'wrecked' immatures after winter gales.

VOICE Grunting at colonies.

CONFUSION SPECIES Similar to Cormorant but smaller and more slender, with a thinner neck, smaller and more rounded headed (with a steeper forehead) and rather thinner bill. Breeding adult distinctive, though crest rapidly lost; in non-breeding birds note colour of face and bill. In flight separated by size, silhouette and more or less straight rather than kinked neck.

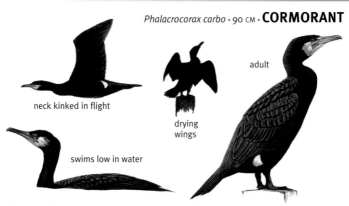

Phalacrocorax carbo • 90 CM • **CORMORANT**

neck kinked in flight

drying wings

adult

swims low in water

A big waterbird common on the coast and larger inland waters. The plumage is not waterproof and it often sits for long periods in a 'heraldic' pose, wings open to dry.

DESCRIPTION Large, with elongated body and neck, long tail and long, straight, hook-tipped bill. Mostly blackish, but cheeks always pale, throat white and bill greyish with bare yellow skin at base. In spring has whitish nape and white thigh-patch; some also have white neck and breast. Sexes similar. Juvenile and immature have variable dirty white underparts. Swims with body low in water, often with head raised; may dive with a jump. Perches upright, on rocks, beaches, trees and even power lines. Flight powerful, with rapid strong flapping interspersed with glides.

POPULATION Common resident.

HABITAT Breeds on rocky shores, in N and W and, increasingly also in trees at inland colonies across Britain. More widespread in winter, both around the coast and in freshwater habitats.

VOICE Deep, harsh calls at colonies.

CONFUSION SPECIES Swimming birds are separated from divers by more angular profile, more hook-tipped bill, often held pointed slightly upwards, and yellowish patch at bill-base. Goose-like in flight, but note long tail. Rather similar to Shag, but larger and bulkier.

immature

juvenile

LITTLE EGRET · 60 CM · *Egretta garzetta*
SPOONBILL · 85 CM · *Platalea leucorodia*

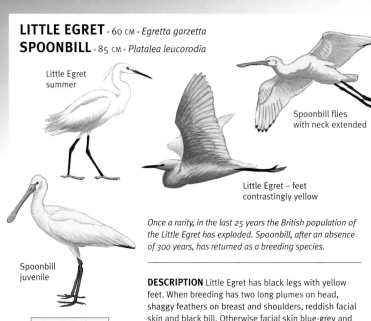

Little Egret
summer

Spoonbill flies
with neck extended

Little Egret – feet
contrastingly yellow

Spoonbill
juvenile

Little Egret

Spoonbill

Once a rarity, in the last 25 years the British population of the Little Egret has exploded. Spoonbill, after an absence of 300 years, has returned as a breeding species.

DESCRIPTION Little Egret has black legs with yellow feet. When breeding has two long plumes on head, shaggy feathers on breast and shoulders, reddish facial skin and black bill. Otherwise facial skin blue-grey and bill duller. Juvenile has dull greenish bill and legs, pinkish bill-base and duller feet. Spoonbill's bill is unique. Adults have short, bushy crest and blackish bill with yellow tip. Immatures have dirty pink bill and legs and black wingtips. Spends long periods loafing; when feeding sweeps bill from side to side to filter food from water.

POPULATION Little Egret is a fairly common resident, with around 1,000 breeding pairs; nests colonially, often with Grey Herons. Typically a handful of Spoonbills wander around the E and S coasts, with largest numbers present in the summer months. In 2010 a Spoonbill colony was established in North Norfolk (8 pairs in 2011).

HABITAT Both are found on estuaries, saltmarshes, lagoons. Little Egret is also increasingly found by rivers and lakes inland.

VOICE Generally silent except at breeding colonies, when both make croaking or grunting noises.

CONFUSION SPECIES None. Spoonbill size as Grey Heron.

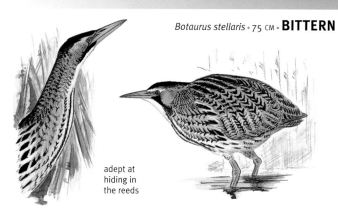

Botaurus stellaris • 75 CM • **BITTERN**

adept at
hiding in
the reeds

*A flagship for wetland conservation, this elusive heron is
on the conservation 'Red List' and is slowly recovering in
numbers from a low point in the 1990s.*

DESCRIPTION Brown, cryptically marked heron with
dark cap, black moustache and blackish streaks on
foreneck. Sexes similar. Juvenile slightly duller. Usually
seen in flight – it is a lucky birdwatcher that sees a
Bittern on the ground.

POPULATION Rare. Due to habitat loss and persecution
the Bittern was extinct as a breeding bird by 1900.
Breeding recommenced and numbers slowly recovered,
but declined again after 1950 to near-extinction by 1997.
Major conservation efforts, especially habitat creation
and improvement, have seen a good recovery, with over
100 booming males by 2011, mostly in S and E England
and at Leighton Moss (Lancashire). Small numbers of
Continental birds winter, mostly in SE England.

HABITAT Reedbeds. Will nest in quite small stands of
reeds, but requires substantial areas nearby for feeding.
In winter disperses to other well-vegetated wetlands.

VOICE Song of male a very low-pitched *wump*, recalling
the sound made by blowing over the neck of a bottle;
this 'booming' can be heard for up to 5km on still nights.

CONFUSION SPECIES In flight may recall an owl or
Buzzard. Compared to Grey Heron wingbeats faster, and
appears broad-winged and thick-necked.

GREY HERON • 94 CM • *Ardea cinerea*

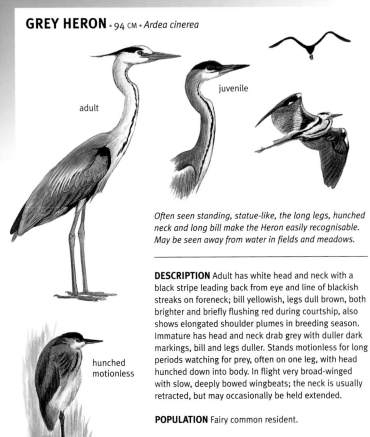

adult

juvenile

hunched
motionless

Often seen standing, statue-like, the long legs, hunched neck and long bill make the Heron easily recognisable. May be seen away from water in fields and meadows.

DESCRIPTION Adult has white head and neck with a black stripe leading back from eye and line of blackish streaks on foreneck; bill yellowish, legs dull brown, both brighter and briefly flushing red during courtship, also shows elongated shoulder plumes in breeding season. Immature has head and neck drab grey with duller dark markings, bill and legs duller. Stands motionless for long periods watching for prey, often on one leg, with head hunched down into body. In flight very broad-winged with slow, deeply bowed wingbeats; the neck is usually retracted, but may occasionally be held extended.

POPULATION Fairy common resident.

HABITAT Fresh or sheltered coastal waters, even visiting garden ponds, as well as wet meadows and pastures. Breeds in large stick nests in tree tops, often colonially; such heronries may be used for many generations.

VOICE Flight call a harsh *frank*.

CONFUSION SPECIES Large size, long neck and greyish plumage rule out everything except Common Crane *Grus grus*. A rare resident in Norfolk, the Crane is larger and holds its long neck extended, has a blackish 'bustle' and a pattern of black, white and red on head.

Pandion haliaetus • 57 CM • **OSPREY**

adult

plunges
feet first to
catch fish

juvenile

*This bird of prey feeds exclusively on fish, hovering
laboriously before diving feet-first into the water. Wiped
out in the 19th century it has made a welcome return.*

DESCRIPTION Larger than Buzzard, wings rather long,
narrowing towards tip and showing only four 'fingers',
tail short and square-cut. When gliding wings held with
a distinct kink at the 'wrist' (carpal) and appear bowed
head-on. White crown and underparts distinctive; adults
have a blackish bar along centre of underwing
separating white coverts from dusky flight feathers;
breast-band averages broader in female. Juvenile has
feathers of upperparts fringed paler, crown streaked,
less boldly marked dark band on centre of underwing
and orange, rather than yellow, eyes.

POPULATION Scarce summer visitor (mid Apr–mid Oct),
winters in Africa. Exterminated in the 19th century but
recolonising Scotland in the 1950s, the Osprey has
slowly increased to 250–300 pairs; returned as a
breeding bird to England (Cumbria) in 2001 and Wales
(Gwynedd) in 2004.

HABITAT Breeds around lakes, rivers and estuaries or
other sheltered coastal waters, nesting on the top of a
nearby tree. On migration found in similar habitats.

VOICE A yelping, whistled *yip-yip-yip*.

CONFUSION SPECIES From afar resembles a large gull,
but note fingered wing-tips and markings on underwing.

COMMON BUZZARD · 54 CM · *Buteo buteo*

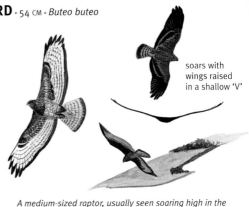

typical individual

soars with wings raised in a shallow 'V'

pale individual

A medium-sized raptor, usually seen soaring high in the sky, giving its characteristic cat-like mewing or sitting hunched on a tree or fence post.

DESCRIPTION At rest appears hunched and 'neckless'. In flight wings broad and blunt, with well-fingered tips, head not projecting far beyond wings, tail short and broad. Plumage very variable. Typically dark above and on breast and underwing coverts, with paler band across lower breast and solid dark patch at carpal joint, underside to flight feathers contrasting paler and greyer, often with a neat dark trailing edge, tail pale with a dark terminal band. Sexes similar. Juvenile as adult but more streaked below and tail lacks broad dark terminal bar. Soars with wings held in a shallow 'V' and tail fanned, glides with wings held flatter.

POPULATION Fairly common resident. Numbers were much reduced by persecution and pesticides; now recovering well, although still relatively scarce in central and E England.

HABITAT Woodland and open country – farmland, moorland or heath.

VOICE A plaintive, slightly nasal *keeoow* (recalling a squeaky toy).

CONFUSION SPECIES Harriers have much longer tail and narrower wings, while Golden Eagle is much bigger, with longer, forward-swept wings and a longer tail.

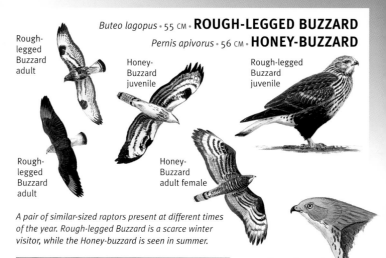

Rough-legged Buzzard adult

Buteo lagopus • 55 CM • **ROUGH-LEGGED BUZZARD**

Pernis apivorus • 56 CM • **HONEY-BUZZARD**

Honey-Buzzard juvenile

Rough-legged Buzzard juvenile

Rough-legged Buzzard adult

Honey-Buzzard adult female

A pair of similar-sized raptors present at different times of the year. Rough-legged Buzzard is a scarce winter visitor, while the Honey-buzzard is seen in summer.

Honey-Buzzard male has blue-grey head

DESCRIPTION Rough-legged Buzzard plumage variable but always contrasting below, with blackish belly, blackish patch at bend of wing, and well defined whitish tail with, in adults, neat dark band at tip (greyer and less contrasting in juveniles). Hovers frequently. Honey-buzzard glides with wings gently downcurved (not kinked), and soars with wings held level. Variable plumage often has parallel dark bars on underwing, two narrow dark bars at base of tail and single black bar at tail-tip.

POPULATION Most years a handful of Rough-legged Buzzards arrive Oct onwards from Scandinavia, mainly on E coast, but there can be influxes of up to 150 in some winters. Honey-buzzard is a summer visitor mid May–Sept with maybe 100 pairs in Britain; winters in Africa.

HABITAT Rough-legged Buzzard favours farmland and coastal marshes, while Honey-buzzard inhabits woodland, including conifer plantations. Generally secretive, it is most likely to be seen when giving its 'sky-dancing' display flight.

VOICE Rough-legged Buzzard has a far-carrying *peee-oooo* call, recalling Common Buzzard. Honey-buzzard is generally silent except in breeding season.

CONFUSION SPECIES Common Buzzard is confusable with both species.

Rough-Legged Buzzard

Honey-Buzzard

RED KITE · 63 CM · *Milvus milvus*

wings held slightly
arched; white flash
at base of primaries
distinctive

*A supremely graceful raptor, until
recently very rare but now the subject
of a successful reintroduction programme.*

tail typically
flexed and twisted
to maintain trim

DESCRIPTION Larger than Buzzard, with proportionally
longer wings and long, clearly forked tail (but looks only
shallowly notched when fully spread). Plumage very
variegated, tail rusty on upperside, more greyish-fawn
below, wings with contrasting pale bar across
upperwings and whitish underside to primaries. Sexes
similar. Juvenile as adult but belly and undertail-coverts
paler. Flight graceful and acrobatic, wings often bent at
wrist. Roosts communally and may flock at feeding
stations.

POPULATION Once common, persecution in the 18th
and 19th centuries led to its near-extinction and
thereafter confined to mid Wales. Due to careful
conservation over 100 years this population, all
descended from one female, grew to 77 pairs by 1991.
Since 1989 reintroduced to the Chilterns, E Midlands,
Yorkshire, NE England and N, central and SW Scotland.
This programme has been very successful and by 2011
around 2,000 pairs were breeding, of which as many
as 900 pairs were in Wales.

HABITAT Mix of farmland with woodland and grassland.

VOICE Various shrill, drawn-out whistles.

CONFUSION SPECIES Unmistakable if seen well.

glides with wings
raised in a shallow 'V'

female

male

*Marsh Harriers are medium-sized raptors with a long tail
and relatively long wings. They fly low over reedbeds,
periods of flapping mingled with long floating glides.*

DESCRIPTION Largest and bulkiest harrier. Males
variegated, wings and tail grey with extensive black
wing-tips and chestnut on upperwing and belly; older
birds paler and may show whitish rump. Female dark
brown with variable creamy crown and forewing and
warm brown tail. Juvenile similar but darker, often
lacking cream on forewing. Soars with wings held in a
shallow 'V', glides with the 'arm' raised but the 'hand'
more level.

POPULATION Once rare indeed, down to a single pair in
1971, has made a welcome comeback and is becoming
ever-commoner in E England, especially East Anglia.
Many are summer visitors (Apr–Oct), but an increasing
proportion (especially females) winter in England.

HABITAT Breeds in extensive wetlands, especially
reedbeds, but forages more widely, often over nearby
farmland, and sometimes nests in crops.

VOICE In display gives squeaky, Lapwing-like *wee-
it...wee-it...*

CONFUSION SPECIES Male may recall male Hen
Harrier, but note chestnut on wings and belly. Female
and immature distinguished from 'ringtail' Hen Harriers
by unbarred tail and lack of white rump.

male

male

MONTAGU'S HARRIER · 45 CM · *Circus pygargus*

male

female

white
rump and
barred tail

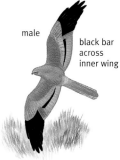

male

black bar
across
inner wing

juvenile

underparts
orange-brown

*Britain's rarest breeding raptor, this summer visitor
is a prize find whenever it is seen. It has been slowly
increasing in numbers, but still merits special protection.*

DESCRIPTION Very similar to Hen Harrier but slimmer,
especially male, with more buoyant, tern-like flight.
Adult males separated by a black bar across secondaries
on both upper- and underwing and tricoloured
upperparts: back and inner wing dark grey, mid wing
pale grey-white, wingtips extensively black. Rusty-brown
spots on underwing-coverts and flanks visible at close
range. Females very similar to Hen Harrier, although
underwings more boldly barred, and only separable with
much experience. Juveniles have distinctive unstreaked
orange-brown underparts.

POPULATION In recent years average breeding
population 15 pairs. Very scarce on passage.

HABITAT Nowadays most nests are on farmland, while
on passage could be seen in any area of open country.

VOICE During aerial courtship display the male gives a
squeaky, nasal *keh-keh-keh-keh*.

CONFUSION SPECIES Hen Harrier.

female

white rump
and barred tail

juvenile

This graceful raptor is still routinely and illegally persecuted in parts of its breeding range and is on the conservation 'Red List'.

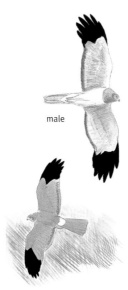

male

DESCRIPTION Rather slender, white-rumped harrier. Male grey with 'dipped in ink' black wing-tips; whitish body contrasts with grey hood. Immature males have some brown in plumage. Females and juveniles, known as 'ringtails', are brown above with a prominently barred tail and underside to flight feathers. Both have streaked underparts, but ground colour of adult female is dingy white, of juvenile rufous-ochre. Flight typical of harrier: long, low floating glides with wings held in shallow 'V'.

POPULATION Scarce resident, breeding in N and W of Britain. In winter, when numbers boosted by Continental immigrants, thinly scattered across Britain and Ireland, with most in coastal regions of SE England and very few in Midlands.

HABITAT Breeds on heather moorland and young forestry plantations. Winters in open country, elusive during the day but gathers to roost communally on saltmarshes and heaths and in reedbeds.

VOICE In aerial display male gives rapid, squeaky *chik-ik-ik....*

CONFUSION SPECIES Lack of rufous on wings or belly separates male from male Marsh Harrier, while 'ringtails' are identified by white rump and barred tail.

GOSHAWK • 48–62 CM • *Accipiter gentilis*

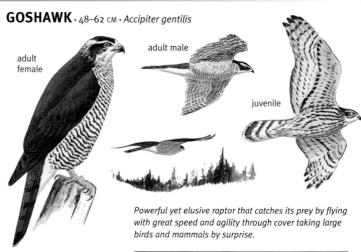

adult female

adult male

juvenile

Powerful yet elusive raptor that catches its prey by flying with great speed and agility through cover taking large birds and mammals by surprise.

adult

DESCRIPTION Male around size of Carrion Crow, female closer to Common Buzzard. Broad-winged like buzzard, but with proportionally much longer tail. Upperparts dark blue-grey in males, slate-grey in females and dark brown in juveniles. Underparts closely barred in adults, but with numerous drop-shaped spots in juveniles.

POPULATION Around 500 pairs. Was (and still is) much-persecuted in Britain and died out by end of 19th century. Current breeding population originates from captive birds. Easiest to see Feb–Mar in soaring display flights.

HABITAT Woodland, both deciduous and coniferous. Will hunt over adjacent open country.

VOICE A very loud, screaming *ki-ki-ki-ki....*

CONFUSION SPECIES Despite size difference, must be very carefully separated from female Sparrowhawk. In level flight wingbeats slower and stiffer, wings longer and broader-based with a more pointed tip, head and neck more protruding, belly and base of tail broader, and tail corners more rounded. Plumages similar, although Goshawk tends to look 'hooded', with bolder white eyebrow; immatures easily identified by spotted rather than barred underparts.

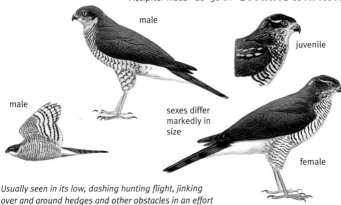

Accipiter nisus • 28–38 CM • **SPARROWHAWK**

male

juvenile

male

sexes differ
markedly in
size

female

*Usually seen in its low, dashing hunting flight, jinking
over and around hedges and other obstacles in an effort
to catch other birds unawares.*

female

DESCRIPTION Male Jackdaw-sized, female rather larger.
Wings short, broad and blunt-tipped, tail long. Male
uniform grey above with 4–5 dark bars on tail, whitish
below with diffuse peachy-orange bars, underside of tail
and flight feathers barred. Female similar but grey-
brown above, off-white below with fine darker bars on
underbody. Juvenile as female but feathers of
upperparts fringed paler, underparts more coarsely
barred. Flight combines rapid, pigeon-like wingbeats
with short glides, also soars, but often unobtrusive and
easy to overlook.

POPULATION Fairly common resident. Pesticide
poisoning led to a huge drop in numbers in the 1950s
and 1960s, but has now fully recovered.

HABITAT Woodland, farmland, parks and gardens,
where sometimes visits bird tables – to feed on the
birds.

VOICE Sharp, staccato *ki'ki'ki'ki'kik*.

CONFUSION SPECIES Separated from Kestrel by
relatively shorter and blunter wings (though difference
not so clear when soaring), many aspects of plumage
and above all by behaviour; never hovers and rarely
perches on wires or telephone poles. See also Cuckoo.

GOLDEN EAGLE • 75–88 CM • *Aquila chrysaetos*

juvenile

soars and glides
with wings raised
in a shallow 'V'

adult

A symbol of majesty and power, the Golden Eagle is confined to the wilder parts of Scotland, with an outpost in the Lake District.

DESCRIPTION Large, wingspan around 1.75 × that of Buzzard (size not always easy to judge), with proportionally long tail (equals width of wing), and much longer, broader wings that are narrow at base with a broad 'hand'. Overall dark brown with golden nape. In flight adult shows paler and greyer base to tail and flight feathers. Juvenile distinct, tail white with neat black terminal band, base of outer flight feathers white. Full adult plumage acquired after 5–7 years. Soars with wings raised in a shallow 'V', often high over ridges; it is a lucky person who sees a Golden Eagle well at close range. Active flight powerful, with 6–7 deep, slow wing-beats followed by a short glide, then more flapping, etc.

POPULATION Around 430 pairs in Scotland. Resident, but immatures may wander.

HABITAT Mountains and moorland. Uses huge traditional nests on cliffs or tree tops.

VOICE Mostly silent.

CONFUSION SPECIES White-tailed Eagle *Haliaeetus albicilla* has been re-introduced to W Scotland. Even larger, but proportionally short tailed (tail all-white in adults), with a long neck and heavy bill. Usually near water.

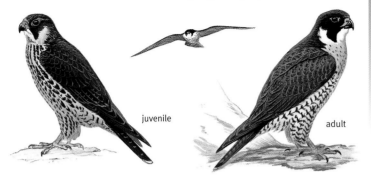

juvenile

adult

The largest resident falcon, hunts birds in high-speed chases, sometimes stooping on prey from a great height, with wings half-closed.

stooping

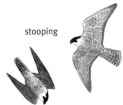

DESCRIPTION Wings relatively broad, tail medium-length with broad base, body broad and powerful. Upperparts slate-grey with dark hood and moustache and contrasting white cheeks. Underparts white, finely barred darker. Sexes similar, but female rather larger. Juvenile browner above, cheeks and underparts washed brown and underparts heavily streaked. Soars, but does not hover; often uses regular perches on rocks and ledges (look for droppings).

adult

POPULATION Uncommon resident. Numbers crashed in the 1960s due to pesticide poisoning, but have now recovered in N and W to all-time high and it is also slowly spreading in lowlands.

HABITAT Mostly wild rocky coasts, mountains and moorland, breeding on cliffs and quarries; in lowlands uses tall buildings, often in urban areas. Largely resident but abandons uplands in winter and disperses to areas where prey concentrates, such as estuaries and other wetlands. Named in the 13th century due to its wanderings or 'peregrinations'.

VOICE A shrill, throaty *khee-khee-khee*.

CONFUSION SPECIES Other falcons are rather smaller and narrower-winged, and lack its bulk and power.

MERLIN · 28 CM · *Falco columbarius*

male

female

female

female

male

male

flight often fast and dashing

Britain's smallest bird of prey, males are barely larger than a Mistle Thrush. This dashing falcon hunts small birds by 'running them down' in flight.

DESCRIPTION Male blue-grey on crown and upperparts, finely streaked blackish, wings and tail similarly blue-grey with black tip to tail and blacker wingtips. Face and underparts dirty white, variably washed buff, with whiter eyebrow and diffuse dark moustache. Female and juvenile with dark brown upperparts, wings and tail boldly barred, underparts off white, heavily blotched blackish. Flight typically fast and low.

POPULATION Over 1,000 pairs breed, almost all in upland areas. Very thinly scattered in winter, despite influx of Icelandic breeders, and best looked for at communal roosts (often with Hen Harriers) in reedbeds, bogs and on heathland.

HABITAT Breeds on moorland. On migration and in winter found in any open country with concentrations of small birds, including farmland, but commonest near coast.

VOICE Call a shrill *kii-kii-kii-kii*.... Only vocal on breeding grounds.

CONFUSION SPECIES Smaller and more compact than Kestrel, never hovers, and never has chestnut-brown, black-blotched upperparts. Female Merlins more likely to be confused with Sparrowhawk, but wings rather more pointed and tail relatively short.

always shows bold white cheeks

adult

Falco subbuteo • 33 CM • **HOBBY**

juvenile

A dashing little falcon that hunts dragonflies as well as swallows, martins and other small birds. Often seen on summer evenings hunting over heathland.

DESCRIPTION Medium-sized with long, rather pointed wings and a medium-length tail. Upperparts dark grey, throat contrastingly white, remainder of underparts appear dark at a distance but in good views bold dark streaks and rufous thighs and vent visible. Sexes similar. Juvenile has pale tips to feathers of upperparts, pale forehead, yellowish-buff underparts and lacks rufous thighs and vent. Active flight fast, with clipped wingbeats; when hunting dragonflies more relaxed, eating them on the wing in short glides. Often hunts in the vicinity of water and active in the evening.

eats dragonflies on the wing

POPULATION Uncommon summer visitor (late Apr–Sept), winters in tropical Africa. Has increased and spread northwards in recent decades.

HABITAT A mixture of woodland, even quite small stands of trees, and open country such as farmland or heathland. Breeds in old nests, especially crows' nests.

VOICE A sharp, clear *kiau-kiau*....

CONFUSION SPECIES Plumage recalls Peregrine, but rather smaller with much more pointed wings. Merlin is also small and dashing, but lacks white cheeks, black moustache and rufous thighs and vent. See also Kestrel.

KESTREL · 34 CM · *Falco tinnunculus*

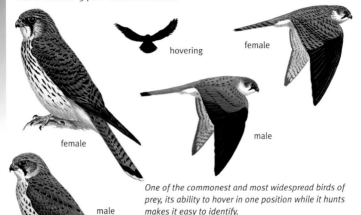

hovering

female

male

female

male

Kestrel

Sparrowhawk

One of the commonest and most widespread birds of prey, its ability to hover in one position while it hunts makes it easy to identify.

DESCRIPTION Medium-sized falcon with long wings and tail; note contrasting dark outer half of wings. Male has blue-grey head, dark moustache, chestnut upperparts and blue-grey tail with a dark terminal band; underparts buff with many dark spots. Female similar but crown and tail browner, upperparts duller brown, barred rather than spotted, and tail barred. Juvenile similar to female but underparts more streaked than spotted. Frequently hovers with the tail depressed, also soars and in active flight wingbeats shallow and rapid with infrequent glides.

POPULATION Fairly common resident, but in winter many abandon upland areas and some migrate to the Continent; conversely, some European birds winter in Britain.

HABITAT Open country, from farmland to heaths and moors, also built-up areas.

VOICE A quarrelsome, piercing *klee-klee-klee*, each note rising in pitch.

CONFUSION SPECIES Sparrowhawk has broad, blunt-tipped wings. Hobby has a proportionally shorter tail and more sharply pointed wings, while Peregrine is much bulkier, with a broader chest and shorter tail.

seldom seen
in flight

juvenile

Shy and secretive, with a narrow body suited to slipping between dense vegetation, this is a hard bird to see well.

white undertail
conspicuous
from behind

DESCRIPTION Smaller than Moorhen, with long legs and long, slightly down-curved bill; tail held cocked and frequently jerked. Adult blue-grey on face and underparts with black and white barring on flanks and conspicuous buff-white undertail; bill mostly red. Sexes similar. Juvenile has shorter, pinker bill and grey on face and underparts replaced by buffy-white with a darker smudge through eye. Small chicks are black. Flies with legs dangled. Keeps to cover, but, especially in early morning and evening, will feed in the open; hard weather also forces it to be more adventurous.

POPULATION Uncommon resident, with some immigration from Europe in winter. Has declined in recent decades due to the loss of its marshy habitats.

HABITAT Dense waterside vegetation, such as reeds and sedges, usually with some open muddy edges; in winter also well-vegetated saltmarshes.

VOICE Varied pig-like squeals and grunts, also an unassuming *kip, kip, kip*....

CONFUSION SPECIES Long bill separates Water Rail from Moorhen. In a rear view, shows complete whitish undertail, whereas Moorhen's is white with a black centre.

MOORHEN · 34 CM · *Gallinula chloropus*

adult

always shows white
line on flanks and
white undertail

juvenile

skittering
on water

*At home on almost any well-vegetated body of fresh
water, this is the commonest and most widespread of
the crake and rail family.*

white
undertail has
black centre

DESCRIPTION Pigeon-sized. Overall dark slate with
back, wings and tail tinged brown. Note yellow-tipped
red bill, red frontal shield on forehead, long greenish
legs and toes, jagged white line along flanks and white
sides to undertail. Sexes similar. Juvenile much duller
mud-brown with paler underparts and a greenish bill,
but already shows whitish on flanks and undertail.
Young chicks are black with a red bill. Usually a little shy,
running or swimming for cover if alarmed, but
nevertheless easy to see. Feeds both on land and on the
water; jerks its cocked tail as it walks, and when
swimming also has a jerky, head-bobbing action.

POPULATION Common resident. Numbers are
supplemented in winter by immigrants from Europe.

HABITAT Well-vegetated bodies of fresh water, from
lakes, reservoirs and gravel pits to very small ponds and
ditches. May be seen feeding on arable land away from
water, particularly in winter.

VOICE Very varied, calls include a mellow, full, explosive
trork.

CONFUSION SPECIES At all ages separated from Coot
by the white line along the flanks and white undertail,
while adults have a red, rather than white, bill.

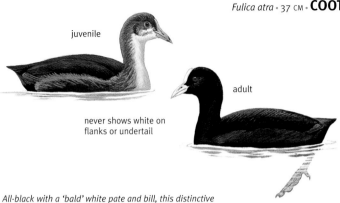

juvenile

never shows white on
flanks or undertail

adult

*All-black with a 'bald' white pate and bill, this distinctive
rail is a familiar bird on larger lakes and reservoirs and
can be found in large flocks in winter.*

wings have narrow
white trailing
edge

DESCRIPTION Larger and plumper than Moorhen,
plumage greyish-black with red eye, white bill, white
frontal shield on forehead and greenish legs and feet
(feet lobed to aid swimming). In flight shows narrow
whitish trailing edge to inner flight feathers. Sexes
similar. Juvenile paler grey with pale face and underparts
and greyish bill. Young chicks black with red and blue on
head and ochre collar. Spends most of time swimming or
loafing on water's edge, but will graze in nearby fields.

POPULATION Common resident. Numbers are boosted
in winter (Oct–Mar) by immigrants from Europe. In
winter forms large flocks, but in summer found in pairs
and very territorial.

feet lobed for more
efficient swimming

HABITAT Lakes, reservoirs and gravel pits, also broad,
slow-flowing rivers and large ponds, but not found on
the small pools and ditches beloved of Moorhen.

VOICE Call a full *kook* and shriller *kit* (given in chorus),
also a clipped *tink*, like metal on stone.

CONFUSION SPECIES Moorhen always shows a whitish
line along the flanks and white sides to the undertail.

OYSTERCATCHER · 43 CM · *Haematopus ostralegus*

summer

boldly black and white in flight

Distinctive, large, black and white wader, common around all coasts and also breeding inland in the N and E, and even on shingled roof tops in some areas.

DESCRIPTION Unmistakable. Adult has bright orange-red bill and eye-ring, deep red eye and pink legs. In winter develops a white 'chinstrap'. Sexes similar. Juvenile and immature duller, more brownish above, with greyish legs, duller eye-ring and eye and extensive dark tip to the more washed-out orange bill; often retains a white 'chinstrap' into its first breeding season.

winter

POPULATION Common breeder, although due to disturbance and development, increasingly local in S England; conversely continues to spread inland in the N and East Anglia. Breeding population resident, although many juveniles move to SW Europe in winter to be replaced by immigrants from Iceland and Scandinavia.

1st-winter

HABITAT Breeds on shingle beaches, dunes, saltmarshes and rocky shores, also inland along river valleys, sometimes away from the immediate vicinity of water in fields and open woodland. In winter much more coastal, with highest numbers on large estuaries.

VOICE Call a shrill *k'leep*, often given in chorus by several birds. Song a repeated *tee-teeoo...*, *tee-teeoo...* given in circling display flight, with exaggerated slow wingbeats.

CONFUSION SPECIES None.

chick

This elegant and distinctive species is the emblem of the RSPB and is now common at many sites on the E and S coasts.

DESCRIPTION Unmistakable. Male has blacker head markings and more gently curved bill, female's head may be tinged brown and the bill is shorter and more abruptly kinked. Juvenile has black areas of plumage brownish and less well-defined, white areas sullied brown, and dull grey legs (pale blue in adult). Feeds by sweeping slightly opened bill from side to side in shallow water. Can swim, and up-ends like a duck.

POPULATION Once common, it was wiped out by hunting, bird collectors and habitat changes by 1840. Recolonised England in the 1940s, initially on the Suffolk coast but then began to spread in the 1970s and now a locally common breeder along the E coast, albeit almost exclusively on nature reserves, from Humberside to Kent. Increasing numbers spend all year on the E coast, but others migrate to SW Europe or Africa in winter. Conversely, European birds join British breeders to winter along coast of SW England.

HABITAT Breeds around shallow brackish lagoons, often with little vegetation, winters on estuaries.

VOICE Call a shrill, yelping *quip quip quip....*

CONFUSION SPECIES None.

STONE-CURLEW • 42 CM • *Burhinus oedicnemus*

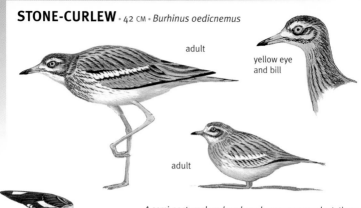

adult

yellow eye
and bill

adult

adult

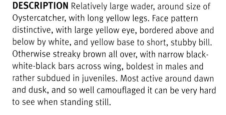

adult

A semi-nocturnal wader whose large eyes are adaptations to help it see in the dark. It is not closely related to a Curlew – the name refers to its Curlew-like call.

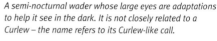

DESCRIPTION Relatively large wader, around size of Oystercatcher, with long yellow legs. Face pattern distinctive, with large yellow eye, bordered above and below by white, and yellow base to short, stubby bill. Otherwise streaky brown all over, with narrow black-white-black bars across wing, boldest in males and rather subdued in juveniles. Most active around dawn and dusk, and so well camouflaged it can be very hard to see when standing still.

POPULATION Around 350 pairs breed in England, mostly around Salisbury Plain and in Breckland of Norfolk and Suffolk. A summer visitor Mar–Sep, but seldom seen on migration. Winters in SW Europe and N Africa.

HABITAT Heathland and arable fields: it prefers mixture of bare, dry, stony ground and short vegetation.

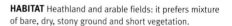

VOICE Song, often given at night, a repeated *krrrr-ee, krrrr-ee...*, with the first note harsh and trilling, the second a shrill whistle; this, and most other calls, are distinctly Curlew-like.

CONFUSION SPECIES None if seen well.

Charadrius morinellus • 21 CM • **DOTTEREL**

summer female

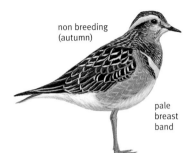

non breeding (autumn)

pale breast band

winter

summer male

The name 'Dotterel' means 'little fool' and is a reference to its tameness. The roles of the sexes are reversed: females are brightly coloured, the male duller.

DESCRIPTION Spring adults distinctive, with bold white line above eye, white throat and grey neck and breast, the latter separated from rusty-orange underparts by a narrow white band. Juveniles and autumn adults duller, with no orange on underparts, but still showing a narrow pale breast band, albeit sometimes faint. In flight no white in wings, rump or tail.

POPULATION Up to 750 pairs nest, most in remote, little visited places. Males incubate the eggs and tend the young, and some females lay 2 clutches, each tended by a different male. Rather scarce on migration, but pairs and small parties ('trips') use stopovers in the lowlands, sometimes traditional.

HABITAT Breeds on bare mountain tops in Scottish Highlands. On passage found on arable fields and short grassland.

VOICE Call purring, song a repeated *phip phip phip phip....*

CONFUSION SPECIES In autumn could be confused with Golden Plover.

LITTLE RINGED PLOVER · 15 CM · *Charadrius dubius*

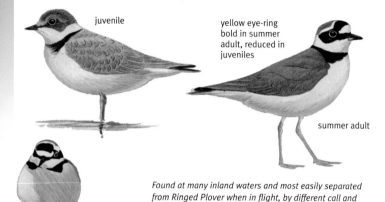

juvenile

yellow eye-ring
bold in summer
adult, reduced in
juveniles

summer adult

winter adult

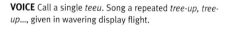

lacks
wing-bar

juvenile

*Found at many inland waters and most easily separated
from Ringed Plover when in flight, by different call and
lack of wing-bars.*

DESCRIPTION Small plover, shows yellow eye-ring in all
plumages. Male has head markings and breast-band
black in breeding plumage, female slightly browner. In
non-breeding plumage both sexes much duller, black
replaced by brownish. Juvenile duller still, with fine pale
feather fringes on upperparts. In flight shows white
sides to tail but no wing-bar; when pressed often prefers
to run rather than fly.

POPULATION Uncommon summer visitor (Apr–Sept),
winters in tropical Africa. First bred in Britain in 1938 and
has slowly spread and increased to around 1,000 pairs.

HABITAT Breeding and on passage around reservoirs,
gravel pits, river shingle, and flooded quarries and
industrial sites; rare on coast.

VOICE Call a single *teeu*. Song a repeated *tree-up, tree-
up...*, given in wavering display flight.

CONFUSION SPECIES Breeding adult separated from
Ringed Plover by eye-ring, kinked rather than rounded
lower border to dark ear-patch, all-black bill and drab
legs. Autumn birds harder as eye-ring reduced and non-
breeding Ringed Plover may have dark bill, but leg
colour still useful. Juveniles have dull eye-ring but note
their small pale forehead and lack of pale eyebrow.

Charadrius hiaticula • 19 CM • **RINGED PLOVER**

juvenile

never shows
pale eye-ring

summer adult

Common on the coast and regular inland, although has declined due to heavy traffic on many beaches. Can only be confused with the rather scarcer Little Ringed Plover.

winter
adult

bold wing-bars

DESCRIPTION In male head markings and breast-band black, in female slightly browner. In non-breeding and juvenile plumage both sexes much duller, black replaced by brownish, and breast-band may be broken in centre. In flight shows white sides to tail and white wing-bar.

POPULATION Fairly common breeding resident. Abundant out of the breeding season with wintering birds from Europe and a spring and autumn passage of Arctic-breeding populations.

HABITAT Breeds on sand and shingle beaches, also inland around gravel pits, river shingle and industrial sites. On passage and in winter in similar habitats but also much more commonly on estuaries and coastal mudflats.

VOICE Call a clearly disyllabic *too-ip*. Song a rapidly repeated whistle, *quiddu-quiddu...*, given in low, stiff-winged display flight.

CONFUSION SPECIES Very like Little Ringed, but has stouter bill and wing-bar and lacks yellowish eye-ring. Breeding birds separated by orange bill-base and orange legs, non-breeders may have dark bill but retain orange legs. Juvenile has legs dull orange to yellowish, but note prominent white eyebrow, lacking in young Little Ringed.

GOLDEN PLOVER · 28 CM · *Pluvialis apricaria*

summer male

winter

narrow white wing-bar

rump dark

white 'armpit'

A wader that is not closely associated with water, breeding mostly on moorland in the N and W and wintering on open fields as often as on mudflats.

DESCRIPTION Large plover; at a distance looks brownish, but at close range note characteristic fine golden spangles on upperparts. Breeding male has blackish face and belly, female less extensively dark. In winter and juvenile plumages sexes similar, lacking black on underparts. In flight shows whitish 'armpits' and narrow wing-bar.

POPULATION Fairly common breeder, though declining, especially when moorland planted with conifers. In winter numbers boosted by immigrants from Europe.

HABITAT Breeds on moors, bogs and rough grassland in uplands, present late Mar–mid Jul. Winters in fields and grassland, both on coast and inland, often in large flocks at traditional sites and frequently with Lapwings. Smaller numbers found on open mud of estuaries, but a scarce visitor to reservoirs and gravel pits.

VOICE Call a sad *pee*. Song plaintive whistles given in butterfly-like display flight.

CONFUSION SPECIES Non-breeding birds recall Dotterel, but Golden Plover is larger and lacks Dotterel's pale breast-band and bold pale eyebrow. See also Grey Plover.

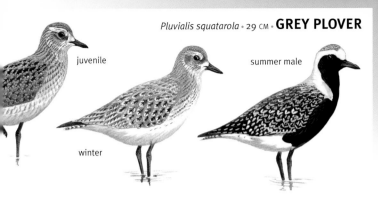

Pluvialis squatarola · 29 CM · **GREY PLOVER**

juvenile

summer male

winter

Has a grey winter plumage and smart black-bellied breeding dress, hence its British and American names, Grey Plover and Black-bellied Plover respectively.

black 'armpit'

DESCRIPTION A large, heavy-billed plover showing a black 'armpit', white wing-bar and white rump in flight. In breeding plumage upperparts boldly spangled silvery-grey, face and underparts solid black in male, more patchy in female. Non-breeding plumage much duller, upperparts brownish-grey with small whitish spangles, underparts pale grey. Juvenile has upperparts spangled yellowish-brown, underparts pale yellowish-buff, finely streaked grey. Typically seen scattered over mudflats, standing still for prolonged periods then taking several steps to pick up food. At high tide roosts communally with Knots etc.

bold white wing-bar

rump white

POPULATION Common passage migrant (late Jul–Nov and Apr–May) with birds en route between high Arctic tundra and SW Europe and W Africa; also common winter visitor.

HABITAT Estuaries and muddy shores, rather scarce inland.

VOICE Call a characteristic plaintive, three-note whistle, *tu-loo-ee*.

CONFUSION SPECIES Non-breeding birds similar to Golden Plover, even more so juveniles, but bigger and bulkier, with heavier bill and, in flight, diagnostic black 'armpit' and white rump.

LAPWING · 30 CM · *Vanellus vanellus*

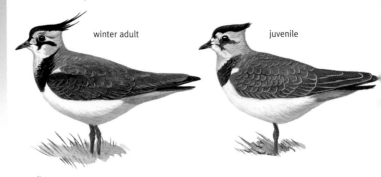

winter adult

juvenile

adult male, summer

A unique crested plover with broad, rounded wings, sometimes called 'peewit' after its distinctive call. From midsummer into winter often occurs in large flocks.

DESCRIPTION Upperparts blackish with green and purple highlights. In summer male has solid black foreneck and face and long crest, female more patchy on face with shorter crest. In winter both sexes have chin and foreneck whitish and fine pale tips to feathers of upperparts; juvenile similar but crest short and feathers of upperparts with complete pale fringes forming faint scallops. In flight slow wingbeats show alternately dark upperwing and white underwing.

POPULATION Common in certain areas, but has declined significantly in recent decades, largely due to a move away from mixed farming that combines arable with pastures. Now commonest in N England and Scotland, but in winter abandons uplands and centre of gravity of population shifts to central and S England, where British birds joined by large numbers of immigrants from Europe.

HABITAT Breeds and winters on arable fields, pastures, meadows and saltmarshes.

VOICE Call a high, nasal *tow-ip* (hence 'peewit'). Song *klep-toowit, klip klip, too-ow-wit*, given in a rolling and tumbling display flight.

CONFUSION SPECIES None.

juvenile

summer female

winter adult

oval white patches at side of rump

summer male; ruff only worn briefly in spring

Unlike other waders the sexes differ markedly in size, males being Redshank-sized, but females, known as Reeves, are not much bigger than a Dunlin.

DESCRIPTION In flight show large white ovals at sides of rump. Male's foppish breeding dress seldom seen in Britain; female merely has a few dark spots on breast. In winter both sexes are plain grey above, pale feather fringes forming a scaly pattern, feathers at bill-base whitish and often shows white eye-ring or even white head; bill may have dull orange base. Juvenile similar but underparts washed peachy, upperparts more boldly scaled with contrasting darker feather centres.

POPULATION Rare and declining summer visitor with a handful breeding in E England. Fairly common on passage, both on the coast and inland, with small but increasing numbers wintering.

HABITAT Breeds in wet meadows. On passage and in winter on coastal pools and lagoons, flooded fields, reservoirs and gravel pits, but seldom open shores.

VOICE Virtually mute.

CONFUSION SPECIES Red-legged males confusable with Redshank, while smaller females may recall several very rare vagrants, but plump body, relatively long neck and small head, combined with short, slightly decurved bill, long olive, yellowish or orange legs and obviously scaly upperparts give it a unique 'jizz'.

DUNLIN · 18 CM · *Calidris alpina*

summer adult

juvenile

winter adult

narrow white wing-bar and white sides to rump

The commonest small wader and very gregarious, often forming large flocks. The original meaning of the name Dunlin 'little brown bird' is apt, but in winter they are greyer.

DESCRIPTION Starling-sized. In breeding plumage cap and upperparts warm brown, breast finely streaked and belly blackish. In winter plumage plain drab grey above and white below. Juvenile brownish above with gingery tone to head and neck, white fringes to feathers of upperparts and black spots on underparts.

POPULATION Fairly common breeder in N and W, though eliminated in some areas by forestry; British and Irish birds winter in W Africa. Common winter visitor to all coasts from N Scandinavia and Russia, additionally birds from Iceland and Greenland pass through on migration. Inland, widespread in small numbers in spring and autumn but scarce in winter.

HABITAT Breeds on wet moorland and coastal grassland. In winter and on passage found on estuaries, muddy shores, coastal pools and grassland, and inland around reservoirs and gravel pits.

VOICE Flight call a harsh *scree*. Song a mixture of calls and prolonged reedy trilling.

CONFUSION SPECIES In juvenile and winter plumage easily confused with other small waders, but note dark centre to rump and tail, medium-length, slightly down-curved bill, and black legs.

Little
Stint
juvenile

Calidris minuta • 13 CM • **LITTLE STINT**
Calidris ferruginea • 19 CM • **CURLEW SANDPIPER**

Little Stint
summer

Little Stint
juvenile

Curlew
Sandpiper
winter

Curlew
Sandpiper
juvenile

*Both these waders occur almost exclusively on
migration, when en route between Africa and their
breeding grounds in the high Arctic.*

Curlew
Sandpiper
summer

DESCRIPTION Little Stint is Britain's smallest wader,
smaller and neater than Dunlin (obvious in direct
comparison), with a shorter bill. Breeding adults rufous
around head and breast. Juveniles obviously scaled
above, with 2 narrow but distinct white lines down back.
Curlew Sandpiper is a little bigger and longer-billed than
Dunlin, in flight its extensive white rump is distinctive. In
summer plumage brick red below. Autumn juveniles
have more neatly scaled upperparts than Dunlin, a
bolder white line above the eye and clean pale
underparts, usually with peachy wash across breast.

POPULATION Both are scarce on passage in May, while
variable number of both Little Stint and Curlew
Sandpiper – from a few dozen to a few hundred,
depending on whether it is a 'good' or 'bad' year – are
recorded in Aug–Sep, mostly on E coast.

HABITAT Coastal lagoons and mudflats, and open
muddy shores of larger gravel pits and reservoirs.

VOICE In flight Little Stint gives a sharp *tit*, Curlew
Sandpiper a chirpy *chirrup*.

CONFUSION SPECIES Dunlin. Spring Little Stints could
also recall Sanderling, while breeding Curlew
Sandpipers recall Knot.

Little Stint

Curlew Sandpiper

95

KNOT · 24 CM · *Calidris canutus*

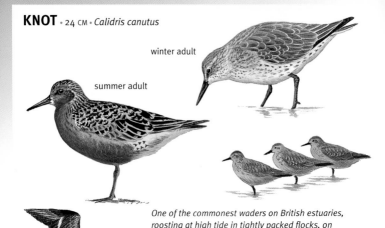

winter adult

summer adult

very nondescript in flight, with narrow white wing-bar and pale grey rump

juvenile

One of the commonest waders on British estuaries, roosting at high tide in tightly packed flocks, on beaches, saltmarshes or even nearby fields.

DESCRIPTION Medium-sized, rotund wader with relatively short, grey-green legs, medium-length bill, grey rump and narrow white wing-bar. Breeding adult rufous-orange below with rusty feathers scattered on upperparts. Winter adult and juveniles very grey with whitish eyebrow, adult has subtle dark streaks on breast and chevrons on flanks, juvenile washed peachy on breast, more spotted than streaked below with fine black scalloping on upperparts.

POPULATION Abundant passage migrant and winter visitor (mid Jul–May). Largest numbers Dec–Mar.

HABITAT Breeds in the high Arctic in Greenland and E Canada, on passage and in winter found on muddy estuaries; rather scarce inland.

VOICE Call a soft, nasal *knut*; also a querulous *quee-quee* in alarm.

CONFUSION SPECIES Very nondescript and best separated from other waders by size, bill length, leg colour and lack of prominent features except white eyebrow. Breeding adult could be confused with Curlew Sandpiper, but that species is more brick-red below with a longer, decurved bill and white rump.

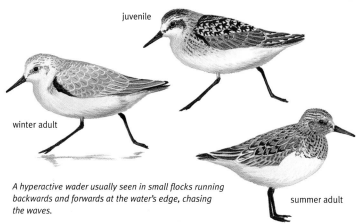
Calidris alba • 21 CM • **SANDERLING**

juvenile

winter adult

A hyperactive wader usually seen in small flocks running backwards and forwards at the water's edge, chasing the waves.

summer adult

DESCRIPTION Size as Dunlin but bill shorter and stouter; bill and legs black. In winter plumage very pale grey above and gleaming white below, depending on posture shows black patch at bend of wing. In fresh breeding plumage (Apr–May) upperparts, head and breast 'salt and pepper' mix of black, white, grey and rufous, becoming more uniformly rufous as summer progresses, especially face and breast. Juvenile spangled blackish above, with touch of yellowish-buff at sides of breast.

bold white wing-bar

POPULATION In coastal areas fairly common in the period mid Jul–May: good numbers winter in Britain and also common on passage to and from wintering grounds in W and S Africa. Scarce inland around lakes and reservoirs, mostly on spring migration in May.

HABITAT Breeds in the high Arctic of Siberia and Greenland, in winter and on passage found on extensive sandy beaches, also mudflats and coastal pools.

VOICE Call a sharp *quick*.

CONFUSION SPECIES Pallid grey winter plumage unique, juveniles and breeding adults can cause confusion, especially away from coastal habitats, and may be mistaken for a stint, but larger, with stouter bill.

PURPLE SANDPIPER · 21 CM · *Calidris maritima*

summer adult

winter adult

juvenile

Almost exclusively confined to rocky shores, including man-made structures such as piers and breakwaters. Can be very tame and reluctant to fly.

DESCRIPTION Slightly larger than Dunlin, very plump with short legs. In winter upperparts, head and breast slate-grey with subtle purple sheen, underparts white with soft grey spots; legs and bill-base contrastingly orange. Breeding adult similarly dark but upperparts a mixture of grey, white and rufous, bill black, legs yellowish-grey. Juvenile similar, but white fringes to feathers of upperparts form bolder scaly pattern. Usually found in small groups but easily overlooked.

dark above with narrow white wing-bars

POPULATION Uncommon winter visitor in the period Jul–May (mostly Oct–Apr) from Norway, Iceland and probably also NE Canada. Most frequent in NE England and Scotland, scarce in SW and S of England; very localised on 'soft' coasts of SE England. A handful of pairs have bred in Scotland since 1978.

HABITAT Winters on extensive areas of seaweed-covered rocks, groynes, etc. Breeds on tundra-like mountain tops, otherwise very rare inland.

VOICE Flight call a conversational *quit*.

CONFUSION SPECIES Adult distinctive, but juvenile can cause confusion, especially in atypical habitat; rather like a dark juvenile Dunlin but note yellowish-orange legs and bill-base.

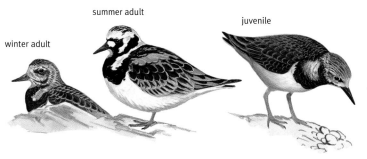

Arenaria interpres • 23 CM • **TURNSTONE**

summer adult

juvenile

winter adult

Named after its habit of flicking over pebbles, seaweed and other strandline debris in search of food. Usually seen in small flocks and can be tame.

summer

winter

DESCRIPTION Larger and stockier than Dunlin with short, stout black bill and short orange legs; in flight shows bold pied pattern. Breeding adult has 'tortoiseshell' chestnut upperparts with neat black and white pattern on head and breast; female slightly duller. Winter adult dark slate-grey above, breast dark with paler patches; juvenile similar but shows paler fringes to feathers of upperparts.

POPULATION Common winter visitor from Canada and Greenland (Aug–May), with Scandinavian breeders passing through in autumn en route to W Africa (they return N via a different route). Small numbers of immatures summer in Britain. Present most of the year.

HABITAT In winter found on estuaries and sandy beaches, but more typically on seaweed-strewn rocky and stony shores. Small numbers occur inland on passage.

VOICE Flight call *tiuw* and a distinctive chattering *kuk-a-kuk-kuk*.

CONFUSION SPECIES Breeding adult unmistakable, in winter and juvenile plumage could be confused with Purple Sandpiper and often in same habitats, but note short, stout bill and variegated head and breast.

WOODCOCK · 34 CM · *Scolopax rusticola*

adult

best seen in roding
display flight at dusk

rises suddenly with
clatter of wings

sitting birds are
superbly camouflaged

This wader breeds inland in damp woodland and, being largely nocturnal, is most easily seen in its characteristic display flight known as 'roding'.

DESCRIPTION Significantly larger than Snipe, with long bill and cryptic coloration. Note transverse bars on head and underparts. Usually solitary and sits tight on ground during the day, where almost impossible to spot, thus often flushed accidentally, appearing large and dark rusty-brown in brief view as it flies off through trees. During Mar–Jul males engage in roding flight, making wide circuits at dusk; large rounded wings, pot belly and long, downward-pointing bill distinctive.

POPULATION Fairly common breeding resident, with much larger numbers of winter visitors from N Europe arriving from Oct onwards.

HABITAT Moist woodland, including young conifer plantations. Feeds at night, probing the ground in wet woodland and, in winter, also in nearby fields. On passage could be encountered almost anywhere, and in hard weather may forage in the open in fields.

VOICE When roding gives a thin *tissik* interspersed with a low grunting *wark-wark-wark* that is only audible as it passes close overhead.

CONFUSION SPECIES Snipe is significantly smaller and paler, usually in different habitats, and calls when flushed.

adult

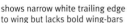

shows narrow white trailing edge to wing but lacks bold wing-bars

Likes to keep to cover and usually seen when accidentally flushed or, more satisfactorily, in its display flight or from a hide overlooking quiet pools.

dives steeply in drumming display

sings from a prominent perch

DESCRIPTION Dumpy, medium-sized wader with short legs and extremely long, straight bill. Plumage brown, cryptically marked, with boldly striped head and white stripes down upperparts. On the ground usually has a crouched posture. Feeds head-down, probing its bill deep into mud. When flushed flies off, calling, in wild zigzag before gaining height.

POPULATION Fairly common breeder, mainly in N and W; has declined greatly in lowlands. Large numbers of European birds winter (Sept–Apr).

HABITAT Breeds on moorland bogs, wet pastures and marshes. On passage and in winter in almost any wet, marshy area with good cover and soft mud, including coastal pools, but abandons upland sites.

VOICE Call when flushed an abrupt harsh *scaap*. Song, often given from a prominent perch, a repeated *chip-er chip-er....* In 'drumming' display flight the modified outer tail feathers make a low throbbing as the bird periodically dives downwards.

CONFUSION SPECIES Jack Snipe *Lymnocryptes minimus*, a scarce winter visitor, is smaller and sits very tight; it usually flushes at one's feet, flies up silently and drops nearby. See also Woodcock.

BLACK-TAILED GODWIT · 42 cm · *Limosa limosa*

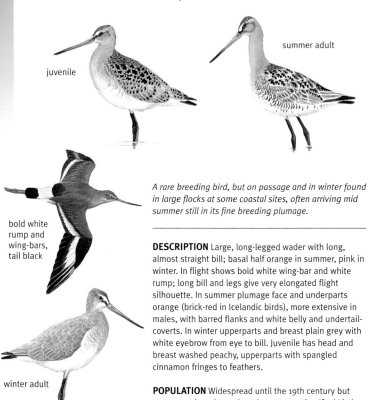

juvenile

summer adult

bold white
rump and
wing-bars,
tail black

winter adult

A rare breeding bird, but on passage and in winter found in large flocks at some coastal sites, often arriving mid summer still in its fine breeding plumage.

DESCRIPTION Large, long-legged wader with long, almost straight bill; basal half orange in summer, pink in winter. In flight shows bold white wing-bar and white rump; long bill and legs give very elongated flight silhouette. In summer plumage face and underparts orange (brick-red in Icelandic birds), more extensive in males, with barred flanks and white belly and undertail-coverts. In winter upperparts and breast plain grey with white eyebrow from eye to bill. Juvenile has head and breast washed peachy, upperparts with spangled cinnamon fringes to feathers.

POPULATION Widespread until the 19th century but now a rare breeder and on the conservation 'Red List', mainly on Ouse and Nene Washes (Cambridgeshire). Fairly common passage migrant and winter visitor (late Jun–May), mostly of the Icelandic form.

HABITAT Breeds on wet meadows. On passage and in winter found on estuaries and coastal lagoons, also flooded meadows in Ireland. Scarce inland in Britain away from breeding sites, but odd birds appears around reservoirs and gravel pits.

VOICE Flight call *wicka-wicka*.

CONFUSION SPECIES See Bar-tailed Godwit.

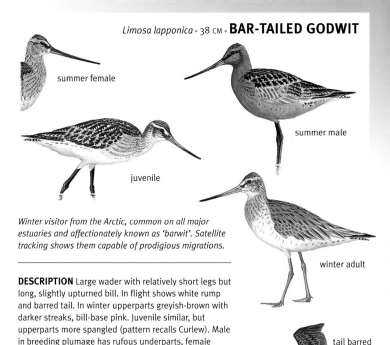

Limosa lapponica • 38 CM • **BAR-TAILED GODWIT**

summer female

summer male

juvenile

winter adult

Winter visitor from the Arctic, common on all major estuaries and affectionately known as 'barwit'. Satellite tracking shows them capable of prodigious migrations.

DESCRIPTION Large wader with relatively short legs but long, slightly upturned bill. In flight shows white rump and barred tail. In winter upperparts greyish-brown with darker streaks, bill-base pink. Juvenile similar, but upperparts more spangled (pattern recalls Curlew). Male in breeding plumage has rufous underparts, female much duller with merely an apricot wash. Gregarious, often seen in large flocks.

POPULATION Common winter visitor (largest numbers Nov–Feb); breeds in arctic Europe and Siberia.

HABITAT Sandy and muddy shores and estuaries. Rather scarce inland.

VOICE Flight call a full but squeaky _kwik-kwik_.

CONFUSION SPECIES Similar to Black-tailed Godwit but has shorter legs (especially above 'knee') and slightly shorter and more distinctly upcurved bill. In winter shows streaked upperparts (Black-tailed is plain grey) and longer white eyebrow, while juveniles are buffy-grey below and appear more streaked above (Black-tailed is peachy below, with distinct cinnamon tone to upperparts, which appear more scaly). Breeding adult has largely black bill, and male has entirely rufous underparts. In flight the two species are very different.

tail barred

Curlew-like in flight, note bill shape

no wing-bars

WHIMBREL · 41 CM · *Numenius phaeopus*

adult

Curlew

Whimbrel

This close relative of the Curlew is usually heard first as it flies overhead with its tittering whistled call, which gave rise to its old country name 'Seven Whistler'.

adult

DESCRIPTION Very like Curlew, but averages around 20 per cent smaller, with shorter bill (although some young male Curlews can be almost as short-billed). Head more boldly marked, with more obvious dark line through eye, more contrasting pale eyebrow, and more solidly dark cap bisected by a thin pale line.

POPULATION Fairly common on spring and autumn passage, with most on the coast and usually just a few, fly-over birds inland. A rare breeding species, with around 300 pairs, and on the conservation 'Red List' due to recent declines.

HABITAT On passage mid Apr–May and Aug–Sep found on mudflats, saltmarshes and coastal grazing marshes. Breeds on moorland in the Northern Isles, principally Shetland.

VOICE In flight gives a very distinctive, soft but penetrating *ti-ti-ti-ti-ti-ti-ti*.

CONFUSION SPECIES Curlew is larger, with a plainer head and longer bill; it never gives a tittering whistle.

Numenius arquata • 50–60 CM • **CURLEW**

bill very long,
especially in
female

The Curlew's call is one of the most evocative bird sounds, both of summer moorlands and bleak winter estuaries and mudflats.

white
wedge
up back

no wing-bar

DESCRIPTION Very large wader with long, strongly down-curved bill. In rather gull-like flight shows dark outerwing and white wedge up back. Overall brown, streaked above and below, and rather plain-headed. Sexes similar, but females have longer bills. Gregarious and often seen in flocks.

POPULATION Common breeder in N and W, but now gone from most of lowlands; many move to SW Britain and Ireland in winter, or to France and Spain. Conversely, large numbers of N European birds winter in Britain.

HABITAT Breeds on moorland, rough grazing and wet meadows. In winter largely coastal, favouring estuaries, mudflats and adjacent fields, with small numbers on inland pastures (and common inland in Ireland).

VOICE Call a haunting whistled *courlee*. Song a beautiful bubbling trill given in display flight.

CONFUSION SPECIES Whimbrel is slightly smaller and shorter-billed with prominent dark stripes on crown, but best identified by call. See also Bar-tailed Godwit.

GREEN SANDPIPER · 23 CM · *Tringa ochropus*
WOOD SANDPIPER · 20 CM · *Tringa glareola*

Green Sandpiper summer

juvenile Green Sandpiper

Green Sandpiper

Wood Sandpiper summer

juvenile Wood Sandpiper

Wood Sandpiper

Green Sandpiper is a fairly common passage migrant. When surprised it flies off calling hysterically and looks like a giant House Martin. Wood Sandpiper is scarcer.

DESCRIPTION Green Sandpiper has the upperparts and breast blackish-brown with scattered small pale spangles (most obvious in juveniles), a pale line from eye to bill and greenish-grey legs. Very black-and-white in flight, with dark underwings and gleaming white rump, but no wingbars. Wood Sandpiper is more slender, with paler, more obviously spangled upperparts, a longer pale line over and behind the eye, and paler and yellower legs. It appears less contrasting in flight, with paler underwings.

POPULATION Green Sandpiper occurs in small numbers, especially in autumn (from July onwards), with a few birds wintering; has bred. Wood Sandpiper is a scarce passage migrant, commonest in May and Aug and near the E coast, and also a rare breeder, with around 20 pairs in Scottish Highlands.

HABITAT Both species favour the margins of reservoirs and lagoons, but Green Sandpiper can also be found on small ponds and ditches and is equally happy in saline conditions, e.g. saltmarsh creeks. Wood Sandpiper is almost always by freshwater.

VOICE In flight Green Sandpipers gives a shrill, loud, ringing, *kluw-ee, kluw-ee-eet….*, and Wood Sandpiper a flat, whistled *chiff-if-if*.

Green Sandpiper

Wood Sandpiper

Actitis hypoleucos • 20 CM • **COMMON SANDPIPER**

juvenile

summer adult

The most ubiquitous of waders, breeding in the uplands of the N and W but occurring on passage around almost any piece of water.

DESCRIPTION Medium-sized sandpiper with mid-length dusky bill and relatively short olive legs. Jizz very distinctive: holds itself horizontally, frequently bobbing its long rear body. In flight wingbeats rapid and shallow, interspersed with glides, and often flies low over water, showing white wing-bars and white sides to tail. Generally olive-brown above and white below, with conspicuous white notch between dark bend of wing and brownish breast. Adults are subtly streaked on upperparts, juveniles have fine scalloping.

POPULATION Fairly common summer visitor and passage migrant (Apr–Sept). Winters in Africa.

HABITAT Breeds along rivers and streams and around lochs and reservoirs, almost exclusively in uplands. On passage around any freshwater, as well as coastal pools and saltmarsh gutters (but seldom open mudflats).

VOICE Call a ringing, high-pitched *swee-wee-wee-wee*, falling slightly in pitch.

CONFUSION SPECIES Flicking flight, call, white wing-bar and white notch at bend of wing unique. Green Sandpiper is larger and darker, lacks the white notch, and has a white rump but no wingbar.

white wing-bar and tail sides

flies low over water

SPOTTED REDSHANK · 30 CM · *Tringa erythropus*

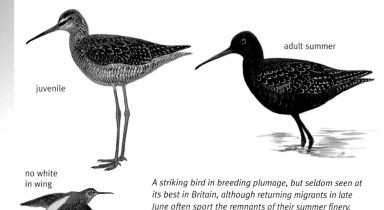

juvenile

adult summer

no white
in wing

winter

winter

moulting

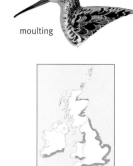

*A striking bird in breeding plumage, but seldom seen at
its best in Britain, although returning migrants in late
June often sport the remnants of their summer finery.*

DESCRIPTION Breeding plumage distinctive. In late
autumn and winter adults clean grey above with
prominent white line above eye and clean white
underparts. Juveniles rather dark, with finely dark-barred
underparts. In all plumages bill long, fine and straight,
reddish towards base, and in flight shows white lozenge
on lower back.

POPULATION Uncommon passage migrant, scarce
inland but may reach double figures on some larger
estuaries. One of the earliest waders to return from its
Arctic breeding grounds in autumn, arriving from late
June onwards. Scarce in winter, with a few dozen around
the Thames in N Kent and Essex, and in Hampshire and
West Wales.

HABITAT Margins of reservoirs and gravels pits, coastal
lagoons and mudflats.

VOICE Flight call a very distinctive 2-note *chew-wit.*

CONFUSION SPECIES See Redshank. Ruff is much
shorter-billed, with narrow pale wingbars and white
horse-shoe on rump.

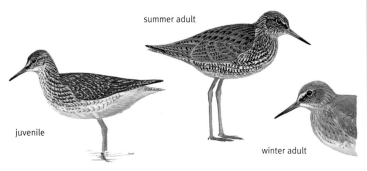

Tringa totanus • 28 CM • **REDSHANK**

summer adult

juvenile

winter adult

Common around the coast in winter and on passage, the red legs and extensive white in the wing make this otherwise nondescript wader easy to identify.

DESCRIPTION Medium-sized with long legs and straight bill. Legs always reddish, bill with variable reddish base. In flight shows white wedge up back and broad white trailing edge to inner wing. Plumage overall grey-brown, breeding adult and juvenile finely streaked and spotted above and below, winter adult rather plainer and greyer. Wary, often the first bird to take flight if disturbed, calling loudly.

POPULATION Fairly common breeder, though has declined in S Britain due to drainage of wetlands and changing agricultural practices. Birds found in Britain are mostly resident, and are supplemented in winter by immigrants from Iceland.

HABITAT Breeds on saltmarshes, wet grassland and in the Hebrides on _machair_. In winter and on passage small numbers occur around reservoirs and gravel pits, but always commonest on estuaries and coastal mudflats.

VOICE Call a mournful, whistled _teu-lu, teu-lulu...._ Song _tu-tu-tu-tu..._ given for prolonged periods in flight.

CONFUSION SPECIES Ruff can show red legs. Spotted Redshank also shows red legs and bill but is larger and longer-billed, and in flight lacks white on the trailing edge of the wing.

broad white trailing edge to wing

breeding birds often perch on posts

GREENSHANK • 32 CM • *Tringa nebularia*

always shows subtly
up-turned bill

summer adult

juvenile

no wing-bars

long white
wedge up
back

winter adult can
look rather pale

The distinctive calls of migrant Greenshanks make them conspicuous on passage and they can sometimes be heard as they fly over at night.

DESCRIPTION Slightly larger than Redshank, bill longer and stouter, subtly but distinctly up-curved, bill-base and legs grey or greenish. In flight shows extensive white wedge up back, wings all-dark. At rest, contrast between dark upperparts and pale belly often striking. Plumage generally brownish-grey above, white below, with a rather pale, finely streaked head and neck. In summer odd blackish feathers scattered in upperparts, head, neck and flanks barred and spotted. Feeds by probing but will also run head-down after small fish. Usually seen singly or in small parties.

POPULATION Scarce breeder, confined to NW Scotland, with some decline due to forestry and disturbance; most Scottish birds winter on coasts of W Britain and Ireland. On passage widespread throughout Britain in small numbers, both inland and on coast, probably mostly birds of European origin.

HABITAT Breeds on bogs, occasionally also in pinewoods. On passage found around gravel pits and reservoirs, coastal lagoons, saltmarsh gutters and estuaries.

VOICE Call a loud, ringing *tew-tew-tew*.

CONFUSION SPECIES Large size, slightly upturned bill, leg colour, flight pattern and call distinctive.

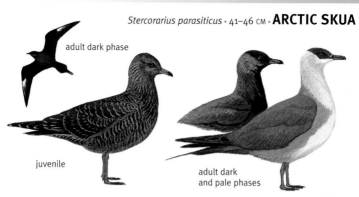

Stercorarius parasiticus • 41–46 CM • **ARCTIC SKUA**

adult dark phase

juvenile

adult dark
and pale phases

*Skuas are piratical seabirds that rob other birds of food.
This is the commonest of the four species of skua that
occur in British waters.*

adult pale
phase

projecting
central tail
feathers

unlike any gull, shows
white flash at base of
flight feathers

juvenile

DESCRIPTION Size as Common Gull; in all plumages
shows whitish flash at base of flight feathers. Adult has
two colour phases, dark and pale, the former uniformly
blackish-brown, the latter with whitish face, neck, breast
and belly, although intermediates occur. In summer
shows projecting central tail feathers. Juvenile finely
barred, but similarly variable, from all-dark to pale
ginger. Long wings give skuas a distinctive rakish
appearance, flight goes from relaxed to fast, agile and
powerful when chasing prey. Readily settles on sea.
Aggressive at breeding colonies, dive-bombing intruders
and will make contact!

POPULATION Scarce breeder in far N of Scotland and
Northern Isles, numbers have declined since the mid
1980s possibly due to food shortages. Regular passage
migrant (May–Jun and Aug–Oct) off all coasts, winters in
S Atlantic. Rare inland.

HABITAT Breeds on coastal moorland, winters at sea.

VOICE Calls mewing, recalling Kittiwake.

CONFUSION SPECIES Three other skuas occur, the
commonest being the larger and more gull-like Great
Skua *S. skua*, which has a similar breeding range. The
identification of the other skuas is often very difficult.

MEDITERRANEAN GULL · 37 CM · *Larus melanocephalus*

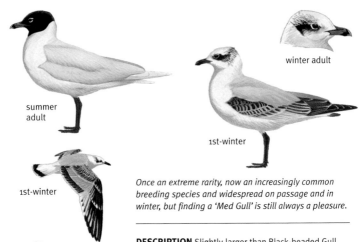

winter adult

summer adult

1st-winter

1st-winter

2nd-winter

summer adult

Once an extreme rarity, now an increasingly common breeding species and widespread on passage and in winter, but finding a 'Med Gull' is still always a pleasure.

DESCRIPTION Slightly larger than Black-headed Gull, with slightly heavier, blunter bill. Bill and legs bright red in adults, blackish in younger immatures.

POPULATION First bred in Britain in 1968 and now over 600 pairs, mostly on S and E coasts. Despite its name, the majority of Mediterranean Gulls breed in the Black Sea. Once thought to be on the road to global extinction, its has increased and spread westwards across Europe in the last 50 years.

HABITAT Breeds around coastal lagoons and spits. More widespread in winter when, although still mainly coastal, scattered birds occur inland, roosting with other gulls on lakes and reservoirs.

VOICE Call a rising, nasal *keeah*.

CONFUSION SPECIES Adults with all-white wingtips can only be confused with much smaller Little Gull. First-winters resemble young Common Gulls, but have more contrastingly blackish and pale grey wing pattern, while older immatures are easily confused with winter/immature Black-headed Gulls, but have more blackish on head and stouter bill.

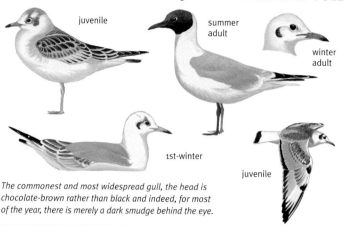

Larus ridibundus · 36 CM · **BLACK-HEADED GULL**

juvenile

summer adult

winter adult

1st-winter

juvenile

The commonest and most widespread gull, the head is chocolate-brown rather than black and indeed, for most of the year, there is merely a dark smudge behind the eye.

DESCRIPTION Small and graceful. Adult has pale grey upperparts with broad white leading edge to outer wing (the best field mark), bordered by dark trailing edge and extensive dark on tip of underwing. In breeding plumage head dark brown, bill and legs blood-red. In winter has dark shadow around eye and dark spot on rear cheeks, bill paler red with dark tip. Juvenile plumage very gingery. First-winter as adult but has brown bands across inner wing, dark tip to tail and paler legs. Very gregarious and often tame, squabbling over scraps.

winter adult

POPULATION Common breeder, mainly in N and W. In winter abundant, although abandons upland areas, with numbers boosted by immigrants from Europe.

1st-winter

HABITAT Breeds colonially on lakes and marshes, both inland and on the coast. In winter ubiquitous: on estuaries, around inland waters, on farmland (where follows the plough), at rubbish tips and in urban areas, where visits gardens. Found almost anywhere inland.

VOICE Calls include a scolding *krreearr* and sharp *kek, kek*.

CONFUSION SPECIES Combination of small size, distinct dark spot behind eye (or in summer, brown hood), reddish bill and legs and white blaze on wing unique.

HERRING GULL · 60 CM · *Larus argentatus*

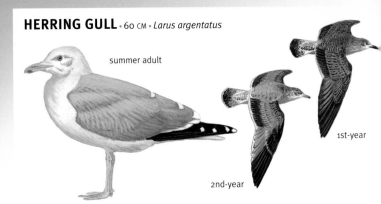

summer adult

1st-year

2nd-year

winter adult

winter adult

A true 'seagull', abundant on the coast and likely to be seen on any visit to the seaside, but also common inland.

DESCRIPTION Large, robust gull. Adult has pale grey upperparts with neat white trailing edge to wing and black wing-tips marked with white 'mirrors'. Bill yellow with a red spot near tip, legs pink. In winter head streaked darker. Juvenile brownish and intricately marked, with black bill. Takes 3–4 years to develop adult plumage and does not show pale grey on upperparts until second winter; until then, hard to separate from young Great and Lesser Black-backed Gulls.

POPULATION Common resident, although few breeding colonies between Yorkshire and Kent, and very widespread in winter, when the population is boosted by large numbers from Europe.

HABITAT Breeds on cliffs, dunes and beaches, also in smaller numbers inland on moorland, sometimes even buildings. In winter attracted to fish quays and rubbish tips as well as more natural sources of food, roosting on the sea, on mudflats and inland on reservoirs and gravel pits.

VOICE Gives the familiar 'laughing' display call.

CONFUSION SPECIES Larger and paler grey than Common Gull, but in immature plumage easily confused with Great, and especially Lesser Black-backed Gulls.

Larus fuscus • 58 CM • LESSER BLACK-BACKED GULL

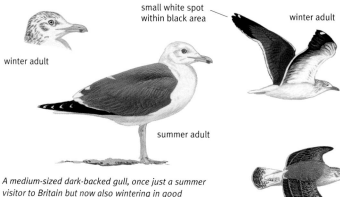

winter adult

small white spot within black area

winter adult

summer adult

A medium-sized dark-backed gull, once just a summer visitor to Britain but now also wintering in good numbers.

DESCRIPTION Adult has slate-grey upperparts and yellow bill with red spot near tip and yellow legs; in winter head smudged grey and legs dull fleshy-yellow. Juvenile relatively dark, with black bill. Takes 3–4 years to develop adult plumage and does not show dark grey on upperparts until second winter, until then hard to separate from young Herring Gull.

POPULATION Common breeder, formerly as a summer visitor but since 1950s has wintered in increasing numbers and now common in winter too, mostly in S Britain, concentrated around good food sources and safe roosts on reservoirs. Small numbers of darker-backed Scandinavian birds also migrate though Britain en route to Africa.

HABITAT Breeds in large colonies on grassland and dunes, both on the coast and inland, also uses flat roofs. In winter widespread, especially inland, around rubbish tips, farmland and pig fields.

VOICE As Herring Gull but deeper and throatier.

CONFUSION SPECIES Dark grey upperparts of older immatures and adults rule out all other gulls except Great Black-backed. Juveniles and younger immatures hard to separate from Herring Gull.

2nd-year

1st-year

GREAT BLACK-BACKED GULL · 64–78 CM · *Larus marinus*

adult

continuous white trailing edge to wing

2nd-year

1st-year

The largest and most impressive gull, standing proud in mixed flocks. More coastal than Herring and Lesser Black-backed Gulls, and generally uncommon inland.

DESCRIPTION Very large with deep, heavy bill; at all ages legs dull pink. Adult has blackish upperparts and yellow bill with red spot near tip. Juvenile has 'salt and pepper' look with pale head and black bill. Takes 3–4 years to develop adult plumage and does not show dark grey on upperparts until second or third winter, until then, hard to separate from young Herring Gull.

POPULATION Uncommon breeder, with most in Northern Isles and absent from E coast of England. Commoner in winter, when resident population supplemented by large numbers from Norway.

HABITAT Breeds singly or in small colonies on islands and rocky headlands with a few inland on moorland. In winter widespread, but large numbers confined to coast and uncommon inland at rubbish tips and reservoirs.

VOICE Deeper, fuller and gruffer than Herring Gull.

CONFUSION SPECIES Bigger and bulkier than Lesser Black-backed with heavier bill. Adults have pink legs, blackish upperparts with a continuous white trailing edge to wing and, in winter, a white head; Lesser Black-backed has legs yellowish, upperparts paler, less white in wing-tip with white trailing edge broken on outer wing, and head smudged grey in winter.

Larus canus • 41 CM • **COMMON GULL**

winter adult

summer adult

1st-winter

adult

2nd-year

1st-year

Breeding in the N and W and largely absent from S Britain in midsummer, this is far from being the commonest gull.

DESCRIPTION Larger and bulkier than Black-headed Gull. In summer plumage adult has white head with dark eye, relatively dark grey upperparts and black wing-tips marked with a large white 'mirror'; bill and legs yellow. In winter head has soft grey streaks and bill duller with dark band near tips. First-winter birds have grey upperparts and midwing panel, remainder of wings brown with extensive dark tips, neat dark band on tail and brownish streaks on head; bill pinkish with black tip, legs pinkish. In second year as adult but more black and less white at wing-tip.

POPULATION Common breeding resident, mostly in N and W. Much more widespread and fairly common in winter, with large numbers of European immigrants.

HABITAT Breeds on bogs, small islands in both fresh and salt water, moorland hills, also man-made sites such as rooftops. In both summer and winter feeds on farmland, especially pastures, and on the coast, and roosts on estuaries, lakes and reservoirs.

VOICE Calls high-pitched, shrill or mewing.

CONFUSION SPECIES Herring Gull is rather similar but much larger and more heavily built, with paler grey upperparts and red spot on bill. See also Kittiwake.

KITTIWAKE · 39 CM · *Rissa tridactyla*

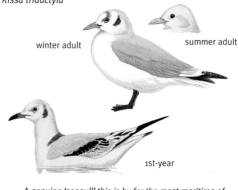

winter adult

summer adult

adult

1st-year

1st-year

A genuine 'seagull' this is by far the most maritime of gulls and away from breeding colonies spends its entire life at sea.

DESCRIPTION Size as Common Gull. In summer plumage adult has white head with dark eye and 'gentle' expression, grey upperparts with a paler outer wings that terminate abruptly in solid black wing-tips, yellowish bill and blackish legs. In winter has grey shawl on nape and dark smudge on rear cheeks. Juvenile and immature have bold black 'W' across wings; attains adult plumage in second or third year, with older immatures resembling a winter adult.

POPULATION Abundant breeder, mostly in N and W due to availability of suitable cliffs. Common migrant in spring and autumn off all coasts, but scarce in winter.

HABITAT Breeds in large colonies on coastal cliffs, sometimes on large buildings. Winters at sea, but small numbers may be seen in inshore waters, especially in rough weather. Rather rare inland, but sometimes visits reservoirs and gravel pits on passage and in winter.

VOICE A nasal *k-kitt-i-waake* at breeding colonies.

CONFUSION SPECIES Adult rather like Common Gull but lacks white spots at wing-tip and has black legs. Immatures share 'W' wing pattern with Little Gull, but are rather larger, with a broad black collar on the hind-neck.

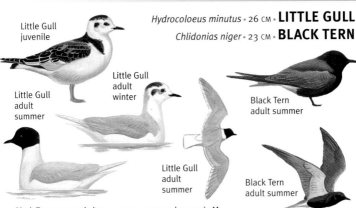

Hydrocoloeus minutus · 26 CM · **LITTLE GULL**
Chlidonias niger · 23 CM · **BLACK TERN**

Little Gull juvenile

Little Gull adult winter

Little Gull adult summer

Little Gull adult summer

Black Tern adult summer

Black Tern adult summer

Black Tern appears in its gorgeous summer plumage in May and in duller juvenile or winter dress in autumn. Little Gull, is Britain's smallest gull and can turn up at almost any time.

Black Tern juvenile

DESCRIPTION Spring adult Black Tern is distinctive. Autumn birds like short-tailed Common Tern, but have blackish hood extending to ear-coverts, dark patch at side of breast, and grey rump and tail. Flight distinctive; light and irregular. Adult Little Gulls have white wingtips, dark underwings (with pale trailing edge), and in summer black hoods. Winter adults have dark cap and spot on ear-coverts, as do first-years, which also show a black 'W' across upperwings in flight.

POPULATION Both are mainly uncommon passage migrants. Black Tern formerly bred in large numbers, but now only rarely breeds. Little Gull nested in 1975 and again a couple of times since then, always unsuccessfully.

HABITAT Both favour lakes, gravel pits, reservoirs and coastal lagoons. Offshore concentrations of Little Gull are sometimes found in N North Sea (late autumn) and Irish Sea (spring).

VOICE Both usually silent; flight call of Black Tern is a harsh *klit* or *klee*, Little Gull a short, hard *kek-kekkek* call.

CONFUSION SPECIES Black Tern in autumn like small Common Tern (see above). Adult Little Gull like Mediterranean Gull, but much smaller with black bill. Immatures resemble young Kittiwakes, but much smaller.

Little Gull

Black Tern

ARCTIC TERN · 34 CM · *Sterna paradisaea*

summer adult

juvenile

no dark
wedge
on outer
wing

distinct
white
trailing
edge to
inner
wing

juvenile

The northern counterpart of Common Tern and hard to separate from it, the Arctic Tern undertakes an epic migration to winter in Antarctica.

DESCRIPTION In summer plumage has long tail-streamers, complete black cap and blood-red bill and legs. Retains summer plumage until arrival in winter quarters. Juvenile subtly barred on upperparts, with dark bar on forewing, white trailing edge to inner wing and black bill. Very aggressive at colonies, will sit on intruders' heads and draw blood with its bill.

POPULATION Common summer visitor (May–Aug), mostly to N and W Britain. Has declined significantly in recent years as lack of food leads to starvation of chicks. Widespread on passage on all coasts, with small numbers migrating overland.

HABITAT Breeds on beaches and coastal grassland; winters at sea.

VOICE As Common Tern but higher and more nasal.

CONFUSION SPECIES Very like Common Tern, but has shorter legs and in summer bill lacks dark tip, tail-streamers longer, extending beyond wing-tips at rest, and white cheeks contrasting subtly with grey-washed breast. In flight upperwing uniform apart from dark trailing edge to outer wing (lacks dark wedge); from below all flight feathers transluscent with neat black trailing edge to outer wing.

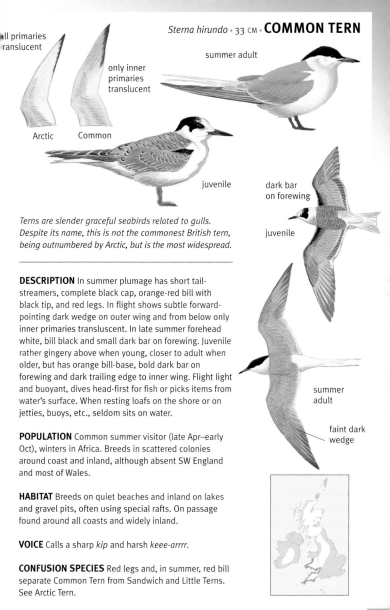

ll primaries
ranslucent

only inner
primaries
translucent

summer adult

Arctic   Common

juvenile

dark bar
on forewing

juvenile

Terns are slender graceful seabirds related to gulls. Despite its name, this is not the commonest British tern, being outnumbered by Arctic, but is the most widespread.

DESCRIPTION In summer plumage has short tail-streamers, complete black cap, orange-red bill with black tip, and red legs. In flight shows subtle forward-pointing dark wedge on outer wing and from below only inner primaries transluscent. In late summer forehead white, bill black and small dark bar on forewing. Juvenile rather gingery above when young, closer to adult when older, but has orange bill-base, bold dark bar on forewing and dark trailing edge to inner wing. Flight light and buoyant, dives head-first for fish or picks items from water's surface. When resting loafs on the shore or on jetties, buoys, etc., seldom sits on water.

POPULATION Common summer visitor (late Apr–early Oct), winters in Africa. Breeds in scattered colonies around coast and inland, although absent SW England and most of Wales.

HABITAT Breeds on quiet beaches and inland on lakes and gravel pits, often using special rafts. On passage found around all coasts and widely inland.

VOICE Calls a sharp *kip* and harsh *keee-arrrr*.

CONFUSION SPECIES Red legs and, in summer, red bill separate Common Tern from Sandwich and Little Terns. See Arctic Tern.

summer
adult

faint dark
wedge

SANDWICH TERN · 39 CM · *Sterna sandvicensis*

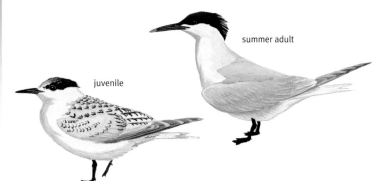

summer adult

juvenile

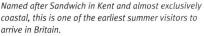

winter adult

adult

Named after Sandwich in Kent and almost exclusively coastal, this is one of the earliest summer visitors to arrive in Britain.

DESCRIPTION Largest breeding tern, with relatively long, narrow wings, short tail and long, slender bill; appears front-heavy in flight. In summer plumage pale grey above, white below with black cap and ragged, squared-off nape; develops a white forehead from Jun onwards. Bill black with small yellow tip, legs black. In flight looks whiter than Common or Arctic Tern, outer flight feathers become increasingly dark as season progresses. Juvenile scaled above with dark cap and short bill. Immature as winter adult but retains some barring on inner wing and has rather dark outer wing.

POPULATION Common summer visitor (late Mar–Sep), winters W Africa. Colonies are scattered round all coasts and even more widespread on migration, but rather scarce inland.

HABITAT Breeds on sand and shingle beaches, often in mixed colonies with other terns and with Black-headed Gulls, dispersing to nearby coastal waters to feed.

VOICE Flight call a grating *kirreet, kirreet....*

CONFUSION SPECIES Larger and paler than other terns, with yellow-tipped black bill, short tail and distinctive call.

juvenile

Sternula albifrons • 23 CM • **LITTLE TERN**

summer adult

winter adult

The smallest tern, breeding in loose colonies on the coast where its favoured sandy beaches are also beloved of holidaymakers; scarce inland.

summer adult

DESCRIPTION Tiny tern with relatively narrow wings and long bill; in silhouette appears front-heavy. In summer plumage outermost flight feathers black, narrow white band on forehead, bill yellow, tipped black, legs orange-yellow. In winter plumage forecrown more extensively white and bill black. Juvenile scaly above with dark bill and dusky forecrown. Flight distinctive, with very fast wingbeats. When feeding hovers, often for long periods, then plunges into water.

POPULATION Uncommon summer visitor (May–mid Sept); winters in Africa. Breeds around all coasts, but with bulk of population in East Anglia. Colonies are vulnerable to flooding and predators, while disturbance and development has concentrated birds into fewer, often more vulnerable, sites.

HABITAT Breeds on sandy and shingle beaches, fishing in nearby inshore waters, estuaries, saltmarsh channels, etc. Occasional on passage inland at gravel pits and reservoirs.

VOICE Calls include distinctive sharp, squeaky *kreet, kreet....*

CONFUSION SPECIES Small size, yellow bill and legs and flight action unique.

GUILLEMOT · 40 cm · *Uria aalge*

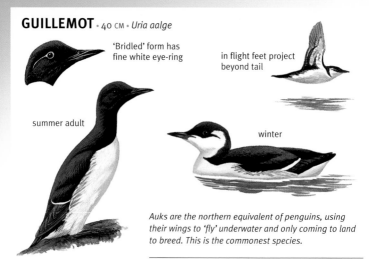

'Bridled' form has fine white eye-ring

in flight feet project beyond tail

summer adult

winter

Auks are the northern equivalent of penguins, using their wings to 'fly' underwater and only coming to land to breed. This is the commonest species.

streaked flanks are diagnostic

DESCRIPTION Bill rather long and slender. In breeding plumage (Nov onwards), upperparts, head and breast dark brown, underparts white, variably streaked on flanks. 'Bridled' variety has thin white eye-ring and is increasingly frequent to N. In non-breeding plumage cheeks white. Sits on sea, bobbing like a cork and diving periodically for food. Like all auks flies low over the water with fast whirring wingbeats, often in straggly lines.

POPULATION Abundant colonial breeder, mostly in the N and W due to the availability of suitable cliffs; no colonies between Flamborough Head (Yorkshire) and Isle of Wight. In winter disperses more widely and can be seen off all coasts, but still commonest near colonies.

HABITAT Breeds on narrow ledges on steep coastal cliffs (pear-shaped eggs won't roll off), dispersing to nearby seas to feed. Rare inland.

VOICE At colonies a rumbling *aarr*.

CONFUSION SPECIES Hard to separate from Razorbill unless seen well enough to judge bill structure, although tends to be paler and browner, with its feet showing beyond tail-tip in flight. If visible, streaked flanks are diagnostic.

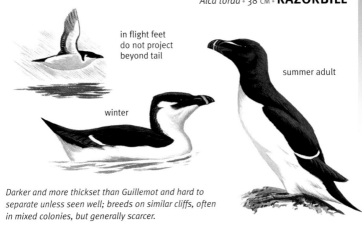

Alca torda • 38 CM • **RAZORBILL**

in flight feet
do not project
beyond tail

summer adult

winter

*Darker and more thickset than Guillemot and hard to
separate unless seen well; breeds on similar cliffs, often
in mixed colonies, but generally scarcer.*

DESCRIPTION Slightly smaller than Guillemot, bill
short, deep and blunt-tipped. In breeding plumage
upperparts, head and breast blackish with pencil-thin
white lines from eye to bill and vertically near bill-tip.
Underparts uniform white. Non-breeding birds have
white cheeks. Immatures lack white line on bill, which is
noticeably smaller and more pointed than adult's. Long
tail often cocked on water and in flight feet do not
project beyond tail-tip.

short
stubby
bill

POPULATION Common breeder, mostly in the N and W.
Usually found at low density, with largest colonies on
boulder screes. In mid Aug–Feb/Mar disperses more
widely and can be seen off all coasts, but still
commonest near colonies and scarce off SE England.

HABITAT Breeds on ledges on steep coastal cliffs,
preferring broader and more sheltered ledges to those
used by Guillemot, also amongst boulder scree and in
rabbit burrows, dispersing to nearby seas to feed. Rare
inland.

VOICE Gives a deep *urrr* at colonies, but rather silent.

CONFUSION SPECIES Hard to separate from Guillemot
unless seen well, especially the smaller-billed
immatures.

BLACK GUILLEMOT · 31 CM · *Cephus grylle*

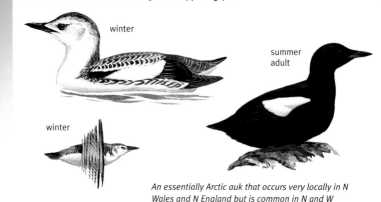

winter

summer
adult

winter

summer

summer

An essentially Arctic auk that occurs very locally in N Wales and N England but is common in N and W Scotland and Ireland.

DESCRIPTION Size between Razorbill and Puffin. In summer plumage black with large white lozenge-shaped patches on wings and red feet; in flight shows white underwings. In winter whitish above with a dusky cap and dark-barred back, wings remain black with white lozenge. Juvenile has sooty crown, nape and upperparts, pale lozenge on wing and whitish underparts, washed brown on neck and barred on flanks. Immature as winter adult but white lozenge spotted with black, sometimes heavily so.

POPULATION Fairly common resident. Rare away from breeding range, the few seen on E and S coasts are probably either immatures or of Scandinavian origin. Extremely rare inland.

HABITAT Breeds singly or in scattered colonies on rocky shores or at the base of sea cliffs, amongst boulders and in rock crevices, dispersing to sheltered coastal waters to feed.

VOICE Displaying male gives a thin pipit-like squeak.

CONFUSION SPECIES Distant flying birds could be mistaken for male Velvet Scoter but have more rapid wingbeats and heavier rear end. In winter plumage may recall a grebe.

winter adult

adult

seldom seen in winter as
usually well out to sea

*A much-loved species, summer adults are instantly
recognisable while the drab non-breeding plumage is
seldom seen as Puffins winter well out to sea.*

dark
underwing

DESCRIPTION Summer plumage unmistakable. In flight
note dark underwings and orange feet. In winter cheeks
dusky grey (head appears all-dark at a distance) and bill
ornamentation duller and reduced in size. Juvenile as
winter adult but bill much smaller and darker. At
colonies usually seen loafing near nest burrows or
bobbing on the sea below the cliffs, with largest
numbers usually present in the evening.

POPULATION Abundant summer visitor (Mar/Apr–Aug),
but very localised and usually only seen around
breeding colonies, which are mostly in N and W, with
none between Flamborough Head (Yorkshire) and
Dorset. Winters well out to sea, most winter records on
the coast refer either to oiled birds or to the occasional
'wreck' when large numbers are driven inshore by
storms.

HABITAT Nests colonially in burrows on steep grassy
slopes around coastal cliffs.

VOICE Various grunting calls given from the nesting
burrow.

CONFUSION SPECIES Juveniles could be confused with
Little Auk *Alle alle*, a scarce winter visitor, but this is
never seen in British waters before Oct.

TURTLE DOVE • 27 CM • *Streptopelia turtur*

adult

tail pattern best seen when tail fanned on take-off or landing

relatively dark in flight

juvenile

A declining species of thickets and overgrown hedgerows, the purring call that gives the species its name is the epitome of high summer.

DESCRIPTION Relatively small, dark pigeon. Adult has rich chestnut-brown scallops on upperparts, blue-grey rump and inner wing, pinkish-grey head and breast and white belly. Note neat patch on side of neck comprising 3–4 thin parallel black and white lines. In flight shows dark underwing and blackish tail with a bold white rim. Juvenile duller and more uniform, lacking neck-patch. Shy and often keeps to cover, but will perch on telephone wires and settle on roads to pick up grit, especially in early morning. Flight fast and agile.

POPULATION Uncommon summer visitor (Apr–Sept), winters in Africa. Has declined greatly in recent decades and now on the conservation 'Red List'. Heavily persecuted on migration in Europe, but agricultural changes in Britain are most likely cause of decline here.

HABITAT Overgrown hedgerows and thickets in arable farmland, bushy rides and clearings in woodland and scrub on heathland and downland.

VOICE A deep, gentle purring *turrrr, turrrr....*

CONFUSION SPECIES Collared Dove has similar white tip to tail but is overall much paler and more uniformly coloured. And, unlike Collared Dove, seldom seen in urban or suburban areas.

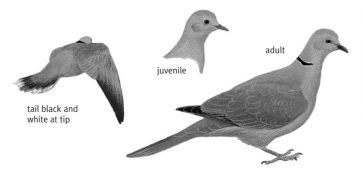

Streptopelia decaocto • 32 CM • **COLLARED DOVE**

juvenile

adult

tail black and white at tip

Colonised Britain as recently as 1955 and now ubiquitous, although always in close association with man.

DESCRIPTION Pale, medium-small pigeon with a relatively long tail. Upperparts fawn with blackish flight feathers and broad white tip to tail, head and underparts pale fawn-grey with narrow black and white half-collar on back of neck. Bill black, legs dull red. Juvenile duller and lacks half-collar. Frequently perches on TV aerials and wires, on alighting raises its tail. Gregarious, often found in flocks where there are plenty of food forms.

POPULATION Common resident. Collared Doves spread rapidly across Europe from about 1930, covering 1,600 km in 20 years, and first bred in Britain in 1955 in Norfolk. Now found throughout Britain and Ireland, being absent only from uplands of N and W.

HABITAT Parks and gardens in both towns and villages, farmyards, grain stores, docks; avoids open country. Often visits bird feeders.

VOICE Calls frequently, a three-note phrase: *who-hooo-ho*, repeated several times; also gives a single harsh purring *kurrr*.

CONFUSION SPECIES Distinctive. In flight, white tip to tail recalls Turtle Dove, but much paler overall, with base of tail fawn rather than black.

pale underwings and broad white tip to tail

Collared Dove

Turtle Dove

dark underwing and narrow white border to tail

STOCK DOVE · 33 CM · *Columba oenas*

adult

distinct black
trailing edge
to wing

nests in holes,
usually in trees

*An unassuming hole-nesting pigeon, often overlooked,
but with a special charm. Once learnt, the neat black
trailing edge to the wing always attracts attention.*

DESCRIPTION Slightly smaller than Feral Pigeon. Head
and upperparts grey, with two short blackish wing-bars
and a glossy green and violet patch on side of neck; eye
dark, bill pink with yellow tip. In flight shows dark
underwings, pale grey rump, tail and wings, a broad black
tip to tail and broad blackish trailing edge to wings.
Juvenile duller than adult, lacking neck patch. Often seen
flying up from quiet country roads in the early morning.

POPULATION Fairly common resident and currently
increasing rapidly, most frequent in regions of mixed and
arable farming. In addition, small numbers of European
birds winter in Britain.

HABITAT Woodland, farmland and parkland with plenty
of mature trees, where nests in tree holes, also some
relatively treeless areas, where nests in cavities in cliffs,
ruined buildings and even rabbit burrows. Will use Barn
Owl or Kestrel nestboxes.

VOICE Call a repeated soft, two-note phrase: *oow-AH,
oow-AH, oow-AH...* becoming more insistent with as the
call goes on.

CONFUSION SPECIES In flight its dark underwing
separates it from Feral Pigeon, and its small size and
lack of white wing-bars from Woodpigeon.

Rock Dove

Feral Pigeon

The familiar street pigeon of towns and cities is descended from Rock Dove but truly wild Rock Dove are now confined to the remote N and W.

striking white underwings

DESCRIPTION Rock Doves have pale grey upperparts with two black bars across the wing, a white rump and white underwing. Feral Pigeons are highly variable, from blackish to white, ginger or piebald, with some identical to Rock Doves.

white rump

POPULATION Rock Doves were domesticated long ago and kept for food. Inevitably some escaped and feral populations became established. Feral Pigeon is now a common resident and indeed has spread and made contact with its wild ancestor, the Rock Dove, and thus populations of wild birds, even in remote areas, are increasingly influenced by Feral Pigeon genes.

HABITAT Feral Pigeons are closely tied to man and live in and around towns and cities, breeding on buildings, but also occur in rural areas where supplies of food such a spilt grain are available. Wild Rock Doves breed in caves and steep cliffs in the N and W.

VOICE The familiar *oow-oooh, oow-oooh....*

CONFUSION SPECIES Feral Pigeons are easily confused with truly domesticated pigeons, such as Racing Pigeons and the various ornamental breeds (all descended from Rock Doves); their separation is often a matter of judgement.

WOODPIGEON · 41 CM · *Columba palumbus*

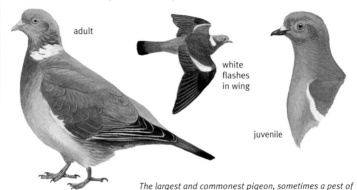

adult

white flashes in wing

juvenile

The largest and commonest pigeon, sometimes a pest of agriculture, with characteristic white markings on the neck and wings.

adult

DESCRIPTION Plump, with a relatively small head. Head and upperparts mid grey with white patch on sides of neck, breast purplish-grey, belly whitish; bill pinkish with a yellow tip, eye yellow. In flight shows broad black tail band, blackish outer wing and grey inner wing separated by a white crescent. Juvenile duller, lacking neck-patch. Frequently on roads and verges, especially in the morning and evening. In display flies upwards with rapid wingbeats, claps wings a few times and then glides downwards; repeats this several times.

POPULATION Abundant resident, with an estimated 4 million pairs in Britain and Ireland. Numbers are boosted in winter by immigrants from Europe, when may occur in very large flocks.

HABITAT Woodland, farmland, parkland and wooded areas of towns and cities, building a flimsy stick nest in trees and foraging on fields, grassland and lawns; increasingly seen around human habitation.

VOICE Song a five-note phrase: *who cooks for-you, oh...*, repeated 3–5 times.

CONFUSION SPECIES Large size, white neck patch and white band on wing distinctive.

Psittacula krameri • 40 CM • **RING-NECKED PARAKEET**

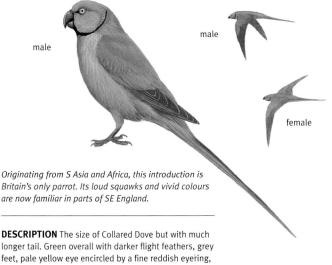

male

male

female

Originating from S Asia and Africa, this introduction is Britain's only parrot. Its loud squawks and vivid colours are now familiar in parts of SE England.

DESCRIPTION The size of Collared Dove but with much longer tail. Green overall with darker flight feathers, grey feet, pale yellow eye encircled by a fine reddish eyering, and red bill. Adult males have longer tail than females, as well as black bib that grades into the narrow pinkish collar. Very noisy in flight, which is fast and level.

POPULATION Presumed to originate from escaped cage birds. First recorded in the wild in Britain in 1969 and now numbering perhaps as many as 30,000 adults, mostly in W London, Surrey, Kent and Sussex (a roost in Esher, used by birds from throughout S London, holds up to 7,000 birds). The increase has been encouraged by warm winters and abundant bird-table food; British birds are now the most northerly breeding parrots in the world.

female

HABITAT Parks, gardens and other areas with scattered trees.

VOICE Call a squawking *kree, kree-kree-kree, kree-kree….*

CONFUSION SPECIES Green Woodpecker has shorter tail, black mask and red crown, as well as very different habits – it never perches in tree tops.

CUCKOO · 33 CM · *Cuculus canorus*

adult

begging nestling

holds
head up in flight

'hepatic' female

Surely the best-known bird song, the Cuckoo's two-note cuck-oo is a sure sign that spring is here. Its breeding biology is, however, rather more sinister.

DESCRIPTION Size as Mistle Thrush but tail rather longer. Male has upperparts, head and breast grey, underparts barred black on white. Bill, eye and feet yellow. Most females as male but with browner barred breast, scarce 'hepatic' females reddish-brown above, strongly barred darker. Juvenile has fine whitish barring on upperparts and pale spot on nape. Flight distinctive, with rapid, shallow wingbeats. Usually sits with wings drooped and tail slightly cocked, sometimes on exposed wires and posts.

POPULATION Uncommon summer visitor (late Apr–Aug), winters in Africa. Has declined in recent years.

HABITAT Occurs wherever its hosts are found together with its principal food, hairy caterpillars – woodland, farmland, heaths, moorland, etc. Cuckoos parasitise a variety of small birds, but especially Meadow Pipit, Dunnock and Reed Warbler. An egg is laid in the nest of the victim and on hatching the young Cuckoo throws out the host's eggs so that it can receive all the attention.

VOICE Song unmistakable, female gives a bubbling trill.

CONFUSION SPECIES Easily mistaken for a Sparrowhawk, but has a slender bill, pointed wings and spots in the tail.

appears ghostly white in flight

adult

Although uncommon and local, this is one of the easiest owls to see where it does occur because it regularly hunts by day.

DESCRIPTION Upperparts a mixture of golden and grey with fine black and white markings, heart-shaped facial disc and underparts white, eyes dark. Sexes similar. Long-winged, looks very pale or white in flight, which is light and wavering, with periods of flapping interspersed with glides and abrupt turns before it drops onto prey. Active at night and during the morning and evening.

POPULATION Uncommon resident, absent from high ground and from much of Ireland and the English Midlands. Declined markedly due to pesticide poisoning in the 1960s and although numbers are recovering, it is now affected by habitat loss and probably also a decline in available nest sites.

HABITAT Farmland, hunting over rough grassland in meadows, pastures, set-aside and road verges, and nesting in old buildings or large tree holes; increasingly uses specially provided nest boxes.

VOICE Relatively quiet. Song an unpleasant, drawn-out grating shriek. Young in the nest give a variety of hisses.

CONFUSION SPECIES None.

chick

135

LITTLE OWL · 22 CM · *Athene noctua*

adult

juvenile

undulating flight

Introduced to Britain in the 19th century, Little Owls are active by day and nest in holes in old trees and sometimes ruined buildings.

DESCRIPTION Small, about the size of a Mistle Thrush but built much more squatly, with a broad, rounded head and short tail. The white eyebrows and yellow eyes give it a fierce expression. Sexes similar. Juvenile duller, with an unspotted crown. Seen singly or in pairs, and in the evening and early morning often sits out on exposed perches in old trees or on telephone poles, fences and buildings. Flight undulating, with rapid bursts of wingbeats interspersed with a pause with the wings closed.

POPULATION First introduced in 1842, then more widely in the 1870s and has now spread to most of lowland Britain. Very rare in Ireland. More recently numbers have declined slightly, perhaps due to habitat loss, especially the loss of suitable nest sites.

HABITAT Farmland, orchards and parkland with plentiful mature trees, old buildings or stone walls.

VOICE Call a sharp yelping *kyow*, falling in pitch, in alarm a rapid series of querulous notes: *chi, chi, chi-chi*. Song a single, plaintive, upwardly inflected hoot (recalling a Curlew), repeated every 5–10 seconds *coo-ee...coo-ee....*

CONFUSION SPECIES None if seen well.

Strix aluco • 38 CM • **TAWNY OWL**

adult

able to fly silently in
search of prey

*Tawny Owls do not hunt by day and are most often seen
in car headlights on quiet country roads or when
mobbed by small birds at their daytime roost.*

DESCRIPTION Plumage an intricate mix of black, white,
brown and grey, forming superb camouflage and making
it hard to see at a daytime roost. Large dark eyes are set
in a rather plain facial disc. Sexes similar. Juveniles leave
the nest before they are fully-feathered and are covered
in fine, grey-brown down. Active at night, but unless
disturbed Tawny Owls do not fly in the daytime.

nestling

POPULATION The commonest and most widespread
owl, although absent from the Isle of Man, most Scottish
islands and Ireland.

HABITAT Woods, well-wooded farmland, parks and
gardens; anywhere with large, mature trees to provide
nest holes and safe roosts; will use nest boxes. Strictly
nocturnal, does not fly by day unless disturbed.

VOICE Calls frequently in winter and early spring, the
commonest call a squeaky *kewick*. Song a drawn-out
hoot followed by a pause and then a rapid series of
notes and another hoot: *hoo-wooh... ho, ho-wo-woo-
wooh*; the female gives a hoarser and more wailing
version. The combination of call and song, cobbled
together in folklore, gives the familiar *tuwit, tuwoo*.

CONFUSION SPECIES Often seen briefly in car
headlights, when appears much darker than Barn Owl.

SHORT-EARED OWL · 38 CM · *Asio flammeus*

juvenile

adult

conspicuous
yellowish patch
near wing-tip

*A day-flying owl of rough country, the 'ears' are, in fact,
tufts of feathers on the head that are rarely visible in
the field.*

streaked breast
contrasts with
unmarked belly

DESCRIPTION Heavily streaked above and on breast
with a plainer and whiter belly. Facial disc pale with dark
circles around yellow eyes, giving a fierce expression. In
flight appears long-winged and blunt-headed. Often
hunts by day, quartering the ground with periods of
slow, loose flapping interspersed with glides; perches
on the ground or on low posts.

POPULATION Scarce breeder, mostly in N and W and
very uncommon in S; absent Ireland. Some migrate to
SW Europe in winter, and there are variable influxes of
European birds into Britain and Ireland from Sept
onwards. Numbers are dependent on rodent
populations, and these are subject to wide fluctuations.

HABITAT Breeds on moorland and in young forestry
plantations, but will use rough grassland, marshes and
dunes. In winter found over any rough open country,
especially near the coast.

VOICE Song a rapid, muffled *bo boo boo-boo-boo...*
given in a circling display flight.

CONFUSION SPECIES Can look very pale in flight, but
never as white as a Barn Owl. Long-eared Owl *A. otus*, a
scarce migrant and breeder, is very similar, but has
orange eyes and, when perched, long 'ears'.

Caprimulgus europaeus • 27 CM • **NIGHTJAR**

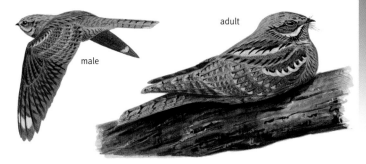

male

adult

A nocturnal species that has gathered much folklore and once had the bizarre name of 'goatsucker'. Nightjars in fact feed on insects caught in flight.

female

DESCRIPTION Thrush-sized, but appears larger in flight, with relatively long wings and tail. Stiff wingbeats are interspersed with long glides and rapid turns; occasionally hovers. Males have white spots on wings and tail, absent in females and immatures. Spends the day resting on the ground, camouflaged by cryptic plumage, and seldom seen unless accidentally disturbed. Becomes active at dusk, but not until light is poor and thus often only seen in silhouette.

active from dusk onwards

POPULATION Uncommon summer visitor (May–early Sept), winters in tropical Africa. Has declined significantly and on conservation 'Red List'.

HABITAT Heathland and conifer plantations, where young plantations and areas of clear-fell provide suitable patches of bare ground.

VOICE Call a full, throaty *cu-wick*, often given in flight. Song a rapid reel (so fast that individual notes can barely be made out), with frequent changes in pitch. Birds sing for long periods from a perch on a tree top or dead limb and the song often ends with a slower, winding down *tow-tow-tow-tow-tow*. Males also clap their wings together in flight to make a snapping noise.

CONFUSION SPECIES None, although a late-flying Kestrel is similar in shape and may cause confusion.

SWIFT · 17 CM · *Apus apus*

has tiny feet adapted to cling to vertical surfaces, but seldom seen perched

A familiar bird in towns and villages, often seen floating high overhead on summer evenings or in tight screaming flocks chasing round the rooftops.

adult

DESCRIPTION Body streamlined, wings long, narrow and sickle-shaped, tail forked (although looks pointed when not spread), plumage blackish-brown with a paler throat. Sexes similar. Juvenile as adult but with narrow paler fringing to its feathers. Normally only ever seen in flight, even mating and sleeping on the wing, and long glides are interspersed with periods of very rapid stiff wingbeats. Can fly very high, but also, especially in damp weather, feeds low, often over water. The tiny feet are adapted to cling to vertical surfaces and Swifts cannot perch on wires or bushes.

POPULATION Common summer visitor (late Apr–Aug), winters in tropical Africa.

HABITAT Breeds in cavities high in buildings, in cities, town and villages (mainly in S and E), dispersing widely to forage. Very rarely, still uses natural nest sites in cliff crevices and also old records of nests in tree holes. Feeds exclusively on aerial insects and small spiders.

VOICE A loud, prolonged and shrill ringing scream, often given in chorus.

CONFUSION SPECIES Swallows and martins are smaller, with relatively shorter, broader wings and a looser flight action; all have pale underparts.

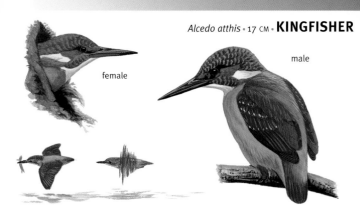

Alcedo atthis • 17 CM • **KINGFISHER**

female

male

Perhaps the most exotic and colourful British bird, yet often elusive and hard to find. Kingfishers are often merely glimpsed as a startling turquoise bullet.

DESCRIPTION A little smaller than a Starling but dumpy, with very short tail and extremely long bill. Male has all-black bill, female shows reddish base to lower mandible. Juvenile as adult but rather duller. Sits motionless low over the water, looking for prey, before plunge-diving, and will also hover when searching for fish. Flight low, fast and straight.

POPULATION Uncommon resident. Largely abandons upland sites in winter when a few move to milder coasts. Very vulnerable to freezing conditions and numbers fall significantly following hard winters.

HABITAT Slow-flowing rivers, lakes, reservoirs and gravel pits, breeds in tunnels excavated in vertical sandy or earthen banks. In winter disperses more widely and found also on seashores and estuaries.

VOICE Call a shrill, penetrating whistle: *zee* or *zee-tee*, often given in flight (louder and more piercing than similar call of Dunnock).

CONFUSION SPECIES None.

plunges
into the
water
head first
to catch
fish

LESSER SPOTTED WOODPECKER · 15 CM · *Dendrocopos minor*

barred back

juvenile

female

male

The scarcest of the three British woodpeckers and always a difficult bird to find; now more than ever following a significant population decline.

DESCRIPTION Sparrow-sized. Black above with white bars across back and wings. Face and underparts white, with black moustache that curls below ear-coverts. Males have red centre to crown, females lack any red in plumage. Generally keeps high in trees and easiest to see in early spring when no leaves and it attracts attention by calling and drumming.

POPULATION Uncommon to scarce, with around 2,200 pairs; four out of five have vanished since the 1970s and the species is now on conservation 'Red List'.

HABITAT Deciduous woodland, old orchards, parkland and stands of trees along rivers and streams.

VOICE Call *kik*, as Great Spotted Woodpecker but weaker. Spring song shrill, squeaky *kee-kee-kee-kee-kee....* In early spring also drums in a very similar fashion to Great Spotted Woodpecker, but each drum-roll softer and longer, at 1.2–1.8 seconds; often gives two drum-rolls in rapid succession.

CONFUSION SPECIES Separated from Great Spotted Woodpecker by barred upperparts without a large white shoulder stripe, and finely streaked underparts, lacking red on belly and vent. Also much smaller.

Dendrocopos major • 23 CM • **GREAT SPOTTED WOODPECKER**

adult female

juvenile

The commonest and most widespread woodpecker, a regular visitor to garden bird feeders. Attracts attention in the early spring by 'drumming' from the tree tops.

DESCRIPTION A little bigger than a Starling, boldly marked black and white, with red vent and undertail-coverts. Male has red patch on hindcrown, female lacks red on head. Juvenile has entire crown red (thus red more extensive than adult male). Flight deeply undulating, with bursts of rapid wingbeats interspersed with periods with wings held closed. Listen for tapping as it explores trees with its chisel-shaped bill.

POPULATION Fairly common resident.

HABITAT Anywhere with large trees, from woodland to hedgerows, parks and gardens, indeed, will even occasionally perch on telephone poles, but needs woodland at least 2–3 ha in extent for breeding.

adult male

VOICE Call a sharp *kik*, sometimes given in slow series. No song, rather in Jan–Jun 'drums' by rapidly beating its bill against a tree to produce a buzzing rattle, 0.4–0.8 seconds long, so fast that individual strikes cannot be distinguished. Pitch and timbre of drumming depends upon type of tree, which acts as a sounding board.

CONFUSION SPECIES Only confusable with the much scarcer, seldom-seen Lesser Spotted Woodpecker, which is sparrow-sized and finely streaked below.

GREEN WOODPECKER · 32 cm · *Picus viridis*

adult female

juvenile

adult male

greenish-yellow
lower back and rump
attract attention

Britain's largest woodpecker, spends most of its time feeding on the ground rather than in trees, as its diet is made up largely of ants.

DESCRIPTION Jackdaw-sized, with greenish-yellow lower back and rump conspicuous in flight. Sexes similar, but at all ages male has red centre to black moustache. Juvenile distinct, heavily spotted on sides of head and underparts and finely flecked on upperparts. Shy, when disturbed flies off to nearby trees, and adept at keeping out of sight behind tree trunks.

POPULATION Fairly common resident, although vulnerable to hard winters. Spreading slowly north in Scotland.

HABITAT Woodland, also well-timbered parkland, farmland, commons and heaths; sometimes visits garden lawns. Breeds in tree cavities and feeds on grubs excavated from rotting wood, but to a large extent also on ants and thus often forages on old established grassland and pasture.

VOICE Call a loud, manic series of 10–20 notes: *kle-kle-kle-kle…*, delivered rapidly but deliberately and falling slightly in pitch towards the end; this call gives it the old country name 'yaffle'. Rarely drums.

CONFUSION SPECIES None, although the yellowish rump sometimes invites confusion with female Golden Oriole *Oriolus oriolus*, a rare migrant and breeder.

adult

black and
white bend
of wing

adult

*Drab plumage conceals one of nature's greatest
songsters – its melancholy song is usually heard as it
sings and circles slowly 100–150m above ground.*

DESCRIPTION Brown and streaky, with fairly obvious
white line above the eye that joins on nape; can show
slight crest. Characteristic black and white patches on
alula and primary coverts at bend of wind. Tail rather
short, tipped white. Feeds on the ground but, unlike
Skylark, will perch on trees and bushes.

POPULATION Very locally common, with recent counts
of over 3,000 pairs in Britain. Present in breeding areas
from Feb/Mar. Wanders more widely in winter, with
some moving to SW England and the Continent.

HABITAT Heathland, especially with bare ground, and
newly clear-felled blocks within conifer plantations,
which also have plenty of disturbed soil.

VOICE Usually sings while circling slowly 100–150 m
above ground. Each phrase is a series of repeated notes,
starting hesitantly but growing in volume and often
falling in pitch: *ti, ti, ti, tuwi-tuwi-tuwi-tuwi-tuwi-TUWI-
TUWI-TUWI...*; after a brief pause a slightly different,
phrase is given, etc. Call a mellow *tit-lwee*.

CONFUSION SPECIES Rather like Skylark, but has
shorter tail without white outer tail-feathers and
distinctive black and white markings at bend of wing.

short tail with
white tip

crest raised

SKYLARK · 19 CM · *Alauda arvensis*

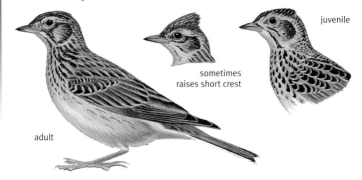

juvenile

sometimes raises short crest

adult

white outer tail feathers and narrow white trailing edge to wing

sings in flight, often high in sky

Famed for its song, the epitome of a summer's day, but sadly in decline since the 1980s and now on the conservation 'Red List'.

DESCRIPTION Slightly bigger than a sparrow, upperparts streaked, underparts off-white with band of streaks across breast. Short crest (in fact more of a squared-off nape), often flattened. In flight note white outer tail feathers and narrow white trailing edge to wing. Sexes similar. Juvenile spotted rather than streaked. Terrestrial, but often hard to see on ground; rarely perches on trees or bushes. Looks relatively large in flight, with broad wings, and characteristically hovers a few feet above the ground before landing.

POPULATION Common but much declined resident, numbers are supplemented in winter by European birds.

HABITAT Any area of open ground including grassland, heathland, moorland, dunes and saltmarsh. Largely abandons upland areas in winter, when often found in flocks.

VOICE Call a musical *tsirrup* or *tsiroeet*. Sings for prolonged periods while flying high in the sky, a continuous and varied outpouring of notes.

CONFUSION SPECIES Separated from sparrows and buntings by more slender, spiky bill and, when visible, the crest, and from pipits by it heavier build, thicker bill and crest. See Woodlark.

adult

spread tail
shows white
spots

juvenile

A familiar species, closely associated with man, feeds over farmland and almost always nests on man-made structures.

DESCRIPTION The classic swallow. Note uniform steely-blue upperparts, dark red forehead and throat bordered by blue-black breast-band (although red face hard to see at any distance), pale underwings and deeply forked tail with long outer tail feathers and white spots visible when tail spread. Sexes similar. Juvenile as adult but upperparts duller, tail-streamers much shorter and face paler and more washed-out.

POPULATION Fairly common summer visitor (mid Apr–Oct), although it has declined significantly in recent years. Winters in southern Africa.

HABITAT Farmland and villages. Feeds on the wing, often over water, around cattle or over ripening corn. Nest an open cup placed on beams or ledges in old farm buildings, porches, culverts, etc. A very few birds breed in natural sites such as caves. In the autumn roosts communally in large reedbeds.

frequently flies low
over ground

VOICE Song a delightful twittering, often given while sitting on wires.

CONFUSION SPECIES Separated from House and Sand Martins by uniform blue-black upperparts, and from Swift by white underparts and long tail-streamers.

SAND MARTIN · 12 CM · *Riparia riparia*

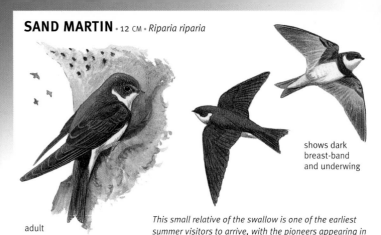

adult

shows dark
breast-band
and underwing

*This small relative of the swallow is one of the earliest
summer visitors to arrive, with the pioneers appearing in
mid March. Usually seen in the vicinity of water.*

juvenile

DESCRIPTION Drab, with brown breast-band but no
gloss to plumage, no tail-streamers and no white in tail.
In flight note dark underwings contrasting with white
belly. Sexes similar. Juvenile as adult but with pale
fringes to feathers of upperparts, throat tinged ochre-
grey and breast-band washed-out.

POPULATION Summer visitor (mostly early Apr–Sept),
wintering in W Africa. Fairly common, although has
declined in recent decades, the population periodically
crashing following droughts in Sahel region of Africa.
Gregarious, and usually seen in flocks.

HABITAT Could be seen anywhere, but breeds colonially
in vertical banks (e.g. river banks, old sand pits, soft
cliffs), tunnelling up to one metre horizontally into the
sand or earth, and usually forages over freshwater. In
autumn may roost in large numbers, together with
Swallows, in extensive reedbeds.

VOICE Distinctive rasping *ree-ree-reep-eep*, draws
attention to its presence.

CONFUSION SPECIES Brown upperparts and breast-
band diagnostic, although they can be difficult to see on
distant flying birds; lacks white rump of House Martin
and long tail of Swallow.

Delichon urbicum • 12.5 CM • **HOUSE MARTIN**

adult

most nests placed
under the eaves
of houses

*A strikingly black and white relative of the swallow, often
siting its mud nest under the eaves of houses. Frequently
seen feeding together with Swifts on summer evenings.*

often flies high

DESCRIPTION Upperparts glossy blue-black with a
startling white rump and white underparts and
underwings. Sexes similar. Juvenile as adult but duller
above with some pale feather fringes and duller white
below. Tends to flight higher than Swallows, attracting
attention with its calls. Often seen on ground, gathering
mud from puddles for its nest.

POPULATION Fairly common summer visitor (late
Apr–Oct), winters in Africa. Often nests in small colonies.

HABITAT Closely associated with towns and villages as
most nests are on buildings, only a small minority
nesting on natural sites such as cliffs. Favours areas
where suitable nest sites are found near to stands of
trees.

VOICE Call a stony *prrit, prri-it, prrit*. Song a jumbled
warble given near the nest.

CONFUSION SPECIES Easily distinguished from Swift,
Swallow and Sand Martin by white rump.

ROCK PIPIT · 16.5 CM · *Anthus petrosus*

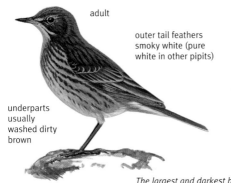

adult

outer tail feathers
smoky white (pure
white in other pipits)

underparts
usually
washed dirty
brown

voice and display flight
very like Meadow Pipit

*The largest and darkest breeding pipit, almost
always seen along rocky coasts and often tame
and approachable.*

'parachuting'
display flight

DESCRIPTION Rather dark overall. Upperparts olive-
brown with diffuse streaks, underparts with smudgy
dark streaks overlaid by a dull wash. Bill black. Sexes
similar, immature as adult.

POPULATION Fairly common resident. On coasts in N
and W, winter visitor to SE England.

HABITAT Breeds along rocky shores. In winter a little
more widespread, visiting saltmarshes and occasionally
inland waters.

VOICE Very like Meadow Pipit. Call a single sharp *phist*
(Meadow Pipit typically gives several notes), alarm a
nervous *tsip*. Song a repletion of simple notes, slightly
fuller and more metallic than Meadow Pipit but similarly
given as it flies up from ground and then 'parachutes'
back down.

CONFUSION SPECIES Separated from Tree and
Meadow Pipits by slightly larger size, drab plumage
(especially underparts), grey-white outer tail feathers
and dark brown legs (pink in Tree and Meadow). Habitat
also a good clue, although Meadow Pipits are frequently
found in clifftop grassland and moorland and will forage
on beaches.

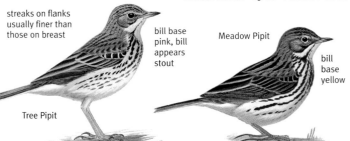

Anthus pratensis · 14.5 CM · **MEADOW PIPIT**

Anthus trivialis · 15 CM · **TREE PIPIT**

streaks on flanks usually finer than those on breast

Tree Pipit

bill base pink, bill appears stout

Meadow Pipit

bill base yellow

Pipits are sparrow-sized, slender, streaked birds with white outer tail feathers that feed on the ground but often perch on bushes and wires.

DESCRIPTION Tree Pipit is slightly bulkier and more conspicuously marked than Meadow, with a bigger, paler bill. Best separated by call and song.

POPULATION Meadow Pipit is a common resident, its numbers supplemented by winter visitors. Tree Pipit is an uncommon and declining summer visitor (mid Apr–Sept); it winters in tropical Africa.

HABITAT Meadow Pipits are found in rough grassland on heaths, moors, pastures and saltmarshes, and in autumn and winter also on arable fields, lakesides and seashores. Tree Pipits breed amongst scattered trees on heaths and downs, in woodland clearings and rides, young conifer plantations and open upland woodlands. On passage habitats as Meadow Pipit.

VOICE Meadow Pipit is very vocal, flight call a thin *see seet*, in alarm a nervous *stit*. Song a series of simple notes, often given as it flies up from ground and then 'parachutes' down, e.g. *chi chi chi su su su su sisisisi tuu tuu tuu*. Tree Pipit calls less often, a single hoarse *spiz*. Song rich, varied and Canary-like, given from a tree top or in a parachuting display flight that starts and ends on an elevated perch.

CONFUSION SPECIES See Rock Pipit and Skylark.

Tree Pipit

Meadow Pipit

Tree Pipit

YELLOW WAGTAIL · 17 CM · *Motacilla flava*

summer male

juvenile sometimes shows only faint trace of yellow

female

A relatively short-tailed wagtail, the breeding male is the yellowest small British bird, rivalled only by Yellowhammer. Often seen feeding around the feet of cattle.

DESCRIPTION Tail relatively short for a wagtail. In all plumages has black legs, blackish tail with conspicuous white outer tail feathers and two narrow pale wing-bars. Spring male has greenish upperparts and bright yellow eyebrow, face and underparts, female and autumn male duller above, paler and more washed-out yellow below. Juvenile duller still, with reduced yellow, although usually still some on undertail-coverts; narrow dark band across lower throat is lost after late summer moult.

POPULATION Uncommon summer visitor (mid Apr–Sept). Numbers have fallen and range has contracted in recent decades. Winters in W Africa.

HABITAT Breeds in lowlands in short, often damp grassland, especially if cattle present; also sometimes on arable land. Migrants occur in wide variety of open grassy habitats, often near water.

VOICE Call a loud, shrill *tsweep*. Song merely the repetition of 2–3 notes, a shrill *tsree-ree*.

CONFUSION SPECIES Spring male distinctive. In other plumages drab olive-brown upperparts (rather than clean grey) distinguishes it from Pied and Grey Wagtails, short tail and black legs are further distinctions from Grey.

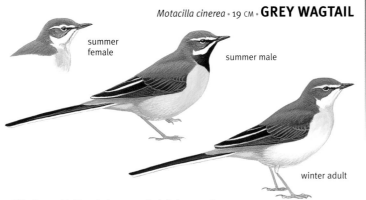

Motacilla cinerea · 19 CM · **GREY WAGTAIL**

summer female

summer male

winter adult

This elegant bird has the longest tail of all the wagtails. Breeding birds are closely associated with water, but more widespread in winter.

DESCRIPTION Always has pale eyebrow, grey upper-parts, yellowish rump, yellow vent and undertail-coverts and fleshy-brown legs. Summer male has black throat, white moustache and yellow underparts, female a faintly darker throat and washed-out yellow underparts. In winter both sexes have a white throat and off-white underparts, tinged yellow on breast. Immatures similar to winter adult, but with stronger pinkish-buff flush to underparts. Constantly pumps its long tail.

POPULATION Fairly common resident, with some immigration from Europe in winter.

HABITAT Strongly associated with water. Breeds along tree-lined rivers and streams, usually fast-flowing and thus mostly absent from lowlands, but will nest around mill-races and weirs. Winters more widely, mostly near water, but also around farm buildings and on rooftops and industrial sites in urban areas.

VOICE Call *tiz-zik*, like Pied Wagtail, but sharper, often followed by single *zik* notes as it flies away. Song a simple *tsi-tsi-tsi* and canary-like warblings.

CONFUSION SPECIES Always separable from Yellow Wagtail by clean grey upperparts, fleshy legs and long tail, and from Pied Wagtail by yellow vent and rump.

white wing-bar visible as it passes overhead

rump yellowish, tail black with conspicuous white sides

PIED WAGTAIL • 18 CM • *Motacilla alba*

long-tailed in all plumages

juvenile

summer male

summer female

Wagtails are sparrow-sized, slender and long tailed. They feed whilst walking or running on the ground, and constantly wag their tails.

DESCRIPTION In all plumages long black tail with white outer tail feathers and two whitish wing-bars are distinctive. Summer male is strikingly black and white, while female has black replaced with dark grey. In winter and immature plumages paler and greyer, but always with some sort of breast-band.

POPULATION Common resident, but absent from higher ground in winter.

winter adult

HABITAT A variety of open habitats, although often associated with water, livestock and/or 'hard' landscape features such as buildings, roads or dry stone walls. Often occurs in towns and cities, on rooftops and car parks (but not gardens unless they have a large lawn or pond). In winter roosts communally, in scrub, reedbeds, at sewage works, etc, sometimes in city centre trees.

VOICE Flight call a sharp *chis-zik*, also gives a more musical *tsu-wee*. Song twittering.

CONFUSION SPECIES Male distinctive; the only other small black and white bird is male Pied Flycatcher, very different in shape, habits and habitat. In other plumages separated from Grey and Yellow Wagtails by lack of any yellow in plumage.

Bombycilla garrulus · 18 CM · **WAXWING**

adult male

rather starling-like
shape in flight

*A strikingly crested, Starling-sized species that is
an irregular winter visitor from Europe, often seen in
urban areas.*

DESCRIPTION The name refers to the red waxy tips to
the inner flight feathers. Female as male but narrower
yellow tip to tail, less well-defined black bib and often
less white on wing-tips (may not form complete 'V's).
Immatures always lack 'Vs', pale fringes merely forming
a whitish line along wing-tip. Often tame and confiding
and usually seen in flocks, sometimes large. Frequently
perches on wires, TV aerials and tree tops.

POPULATION Winter visitor from N Europe (Oct–Apr).
Almost regular E Scotland and NE England, but numbers
very variable. Most winters only a handful arrive in
Britain, but every few years an 'invasion' occurs when
much larger flocks appear following failure of rowan
berry crop in Europe.

HABITAT Feeds on small fruit (as well as flying insects)
and can occur wherever suitable shrubs and trees are
found – supermarket car parks are often favoured.
Flocks are nomadic, moving around in search of food.

VOICE Call a rapid, sibilant trill: *sililililili...*, often given in
chorus.

CONFUSION SPECIES None if seen well, but in size,
shape and flight silhouette close to Starling and easily
overlooked in dull winter light.

DIPPER • 18 CM • *Cinclus cinclus*

adult

juvenile

perches on
boulders and
bobs

*Always found along rivers and streams, flying from
boulder to boulder and plunging into the water to feed.
using its wings to 'swim' underwater.*

DESCRIPTION Starling-sized, but very rotund with a
short tail (often cocked); frequently bobs, flight fast and
direct. Sexes similar. Juvenile greyish above, whitish
below, with many darker bars. Resident birds have a
chestnut belly, immigrants from Europe a blackish-brown
belly.

POPULATION Uncommon resident that has declined
due to pollution and acid run-off from conifers planted in
river catchments. Breeds mostly in N and W and rare in S
and E of England, although a tiny handful of birds from
Europe winter in eastern counties.

HABITAT Rocky rivers and fast-flowing streams, often in
or near woodland. Found mostly in uplands of N and W,
but also sometimes in adjacent lowlands around mill-
races, weirs, etc. Avoids acidic waters. Mostly sedentary,
but some dispersal in winter in hard weather.

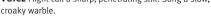

flight fast and low
over water

VOICE Flight call a sharp, penetrating *stik*. Song a slow,
croaky warble.

CONFUSION SPECIES Aquatic habitats, behaviour and
extensive white bib unique.

Troglodytes troglodytes • 9.5 CM • **WREN**

flight low, fast
and direct

adult

One of Britain's smallest birds and a familiar character in folklore, with an astonishingly loud song. Roosts communally – up to nearly 100 birds – in cold weather.

DESCRIPTION Tiny, the short tail is almost always cocked upwards. Overall rich brown, finely barred blackish, with distinct pale eyebrow. Sexes and ages alike. Moves around jerkily in dense cover, flying rapidly from one patch to another.

POPULATION Common resident. In winter withdraws from some upland areas and some immigration from Europe. Affected by long periods of freezing weather, after which numbers may fall sharply.

HABITAT Dense, scrubby vegetation, in woodland, hedgerows, parks and gardens, as well as scrubby valleys on moorland or even exposed sea cliffs and treeless offshore islands. In winter also reedbeds.

VOICE Call a loud *chak-chak-ak*, sometimes extended into a rattle, also a more buzzing *trrrrrrrrrrr*. Song, given from cover, a loud, fast warbling and trilling, lasting 3–6 seconds.

CONFUSION SPECIES Goldcrest is even smaller, but is greenish above, lacks the cocked tail and has yellow stripe on crown. May recall a small warbler such as Chiffchaff if seen briefly in the undergrowth, but unmistakable if seen well.

found in almost any dense, scrubby vegetation

DUNNOCK • 14.5 CM • *Prunella modularis*

adult

frequently seen in groups
of two or three

An superficially sparrow-like bird that is found in many gardens. It is a regular visitor to bird feeders, but often picks up fallen scraps on the ground.

juvenile

sparrow-like but
note fine bill and
blue-grey underparts

DESCRIPTION Robin-sized with a fine bill. Plumage rather dark, generally sparrow-like but sides of head, throat and breast blue-grey. Sexes similar. Juvenile as adult but more boldly streaked on head and underparts. Quiet and inconspicuous, although often tame, usually seen shuffling about on the ground, constantly flicking its wings. Occurs singly or in small groups, never in flocks.

POPULATION Common but declining resident. (One of the most important hosts for Cuckoos, and those that specialise in parasitising Dunnocks have evolved eggs that closely match their unwitting host.)

HABITAT Any area of low, thick scrubby vegetation, in woods, mature hedgerows, heaths, commons, parks and gardens.

VOICE Call a sharp, penetrating *stiih*. Song a sweet whistled ditty, varying little between individuals, *tiddle-iddle-lu-wi, tiddle-iddle-lu-wi-lu-weet*. Usually sings from an exposed perch.

CONFUSION SPECIES Separated from female House Sparrow by very fine bill, slimmer build and behaviour. Often overlooked as a female Robin.

Erithacus rubecula • 14 CM • **ROBIN**

juvenile

adult

Generally regarded as the UK's favourite bird and familiar to everyone, at least on Christmas cards. Highly territorial, Robins sing throughout the year.

DESCRIPTION Orange-red covers both face and breast. Sexes similar. Juveniles lack red; throat and breast have ochre spots and darker scales and the upperparts are also spotted ochre. Within a few weeks, however, they moult into immature plumage, which is very like adult.

POPULATION Common resident. Densities are lowest in the far north and in upland regions, which may be abandoned in winter. A small proportion of British birds winter on the Continent, whilst European birds pass through Britain on migration to winter quarters in SW Europe and N Africa.

HABITAT Woods, parks, gardens; anywhere with some dense shrubby vegetation and preferably at least a few trees. In winter uses an even wider range of habitats, including reedbeds and a regular visitor to bird feeders. Nests in a variety of sites, including nestboxes and shelves and ledges in garages and sheds.

VOICE Call a sharp *tic*, and in alarm a very high-pitched, hissing *tseee*. Song beautiful, 'wistful' and melodic, with no two verses the same. Sings all year, autumn song being the most melancholy. Sometimes sings at night.

CONFUSION SPECIES None. Combination of a red face and breast and fine, insect-eater's bill is unique.

COMMON REDSTART · 14 CM · *Phoenicurus phoenicurus*
BLACK REDSTART · 14 CM · *Phoenicurus ochruros*

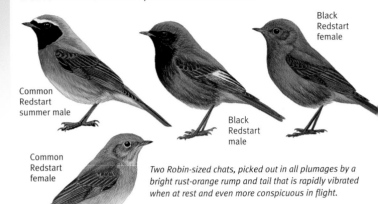

Black Redstart female

Common Redstart summer male

Black Redstart male

Common Redstart female

Two Robin-sized chats, picked out in all plumages by a bright rust-orange rump and tail that is rapidly vibrated when at rest and even more conspicuous in flight.

Common Redstart

Black Redstart

DESCRIPTION Robin-sized. Males distinctive, although in autumn the bright colours of Common Redstart are muted. Female Common Redstart is warm brown above with a narrow pale eye-ring and warm buffy-white below. Female Black Redstart is cold brownish-grey above and mousy grey below. Immature of both species as female, but young males may show traces of adult pattern.

POPULATION Common Redstart is a summer visitor (late Apr–early Oct), winters in Africa. Fairly common in the N and W, but has declined in lowland areas and now gone from much of SE England. Black Redstart is a rare breeding bird in England, with around 50–100 pairs. More widespread on passage on the coast, especially in late autumn and winter.

HABITAT Common Redstart breeds in woodland, parkland, old hedgerows and along wooded moorland streams; requires plenty of nest holes in trees or dry-stone walls. Black Redstart breeds in industrial or derelict urban areas, power stations and cliffs.

VOICE Common Redstart's call is a whistled *huweet*, often followed by *chik-chik*; song a throaty warbler, often incorporating a repeated note or rattle followed by a variety of terminal phrases. Black Redstart's song is a mixture of whistled rattles and strange hissing or fizzing notes.

Luscinia megarhynchos • 16.5 CM • **NIGHTINGALE**

juvenile

usually
sings
from
cover

adult

Closely related to the Robin, this famous songster is much drabber in appearance and much harder to see, as it usually keeps to dense cover.

DESCRIPTION Robin-sized. Warm brown above with rusty rump and tail, dark 'beady' eye set in plain face, whitish throat and pale grey underparts. Sexes similar. Juvenile spotted with buff. Shy, keeping to dense cover and seldom seen when not singing.

POPULATION Summer visitor (late Apr–Aug), winters in Africa. Uncommon and declining, the range is also contracting.

HABITAT Dense thickets in and around woodland and on heaths and downland, overgrown hedgerows, scrub around gravel pits. Has a preference for coppiced woodland and often commonest near the sea.

in flight shows rusty
rump and tail

VOICE Call a Chaffinch-like *heep* and bizarre croaking *errrr*. Song rich, full and powerful, composed of phrases 2–4 seconds long with equally long gaps between them. Each phrase comprises impressive rattles and trills interspersed with a slower, pure, whistled crescendo: *tu-tu-tu-tu-tu...*. Sings by day as well as by night, but most persistent between midnight and dawn. Song period short, mid Apr–May, with little song later in summer.

CONFUSION SPECIES Unique if seen well. Juvenile spotted like juvenile Robin but still has contrasting rusty tail. See also female Redstart.

WHINCHAT · 12.5 CM · *Saxicola rubetra*

autumn adult / 1st-winter

summer male

female

A very pretty chat that has vanished as a breeding bird from much of England, although still widespread on migration.

in flight shows white at base of tail

DESCRIPTION Robin-sized, usually sits upright on bushes, fences, wires and other prominent perches, flicking its wings and tail. In all plumages has prominent pale eyebrow, apricot-washed breast, boldly streaked rump and some white at base of black tail. Spring male brightest, with blackish cheeks set off by white eyebrow and moustachial stripe and white patch in wing. Female and immature duller lacking white in wing.

POPULATION Fairly common summer visitor (May–Sept), winters in Africa. Has declined significantly and now absent from most of lowland England.

HABITAT A mixture of rough, often damp grassland with bushes or other suitable perches, e.g. heathland, grassy moorland, young conifer plantations, sometimes bracken-covered slopes. More widespread on migration, in any open bushy habitat.

VOICE Call a piping *hip*, often combined with *tak-tak*. Song a short, fast, rich but scratchy warble with distinct pauses between phrases.

CONFUSION SPECIES Shape and habits recall Stonechat and Wheatear, but always separated by combination of pale eyebrow and streaked upperparts and rump.

juvenile

female

summer male

An attractive, perky chat that sits up on bushes, flicks its wings and tail and gives a call that sounds like stones being knocked together, hence its name.

autumn female

DESCRIPTION Robin-sized. In all plumages tail uniformly dark, long but narrow white patch on inner edge of folded wing (often hidden), underparts peachy-orange, head lacks pale eyebrow. Spring male has sooty head, broad white half-collar and very dark upperparts. Female similar, but paler with head pattern much more subdued. In autumn male's plumage muted by pale feather fringes and female often lacks dark throat. Just out of nest juvenile dark and streaky, but after late summer moult resembles autumn female.

POPULATION Fairly common resident, but vulnerable to cold weather and in winter many desert uplands and E Britain for milder S and W, with some moving to Europe. Typically uses a prominent perch; usually seen in pairs.

in flight shows slightly paler rump and white shoulder patches

HABITAT A mosaic of bushes, scrub and more open areas, especially heathland and moorland; often commonest near coast.

VOICE Call a high-pitched *vist*, often followed by a hard *chak-chak*. Song short, like a abbreviated Dunnock, sometimes given in brief song flight.

CONFUSION SPECIES Similar to Whinchat, but tends to be darker and more uniform, lacking pale eyebrow and white at base of tail.

WHEATEAR · 15 cm · *Oenanthe oenanthe*

summer male

female

autumn adult / 1st-winter

One of the world's great migrants. Breeds from Greenland right across Europe and Asia to Alaska, the entire population wintering in tropical Africa.

juvenile

in flight note white rump and tail with distinctive inverted 'T'

DESCRIPTION Robin-sized. In all plumages shows inverted black 'T' on tail surrounded by bold white rump and tail sides (the white 'arse' that gives it its name), pale eyebrow and buffy-white underparts. Spring male grey above with black bandit-mask and black wings, female duller, with browner wings and no mask. Autumn birds hard to age, generally brownish above, darker wings of males obscured by paler fringes. Juvenile dark, scaly and spotted. Usually feeds on ground, but will perch on low bushes, rocks, wires, etc.

POPULATION Fairly common summer visitor (late Mar–Sept). Breeds mostly in uplands of N and W and has declined in lowlands, where now largely coastal in breeding season. Widespread on passage.

HABITAT Open areas with short, grazed grassland and holes for nesting, either cavities in drystone walls and rock piles, or rabbit burrows. Can occur in almost any open grassy area on migration.

VOICE Call a piping *hip*, often combined with a hard *chak-chak*. Song a short, scratchy warble, sometimes given in flight.

CONFUSION SPECIES Black and white pattern of rump and tail unique.

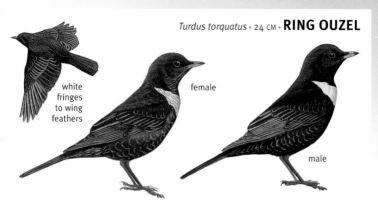

Turdus torquatus • 24 CM • **RING OUZEL**

white fringes to wing feathers

female

male

The upland counterpart of Blackbird, fussy about its habitat during breeding season but much more widespread on migration when it can turn up anywhere.

juvenile

DESCRIPTION Adult male blackish with white half collar and fine whitish fringes to wing feathers. Female tends to be slightly less black, with breast band washed brownish. Immatures have, at most, faint 'shadow' collar, fine pale fringes to wing feathers, and faint pale edges to feathers of underparts, giving subtly scaled pattern.

POPULATION Uncommon and local summer visitor and passage migrant (mid Mar–Oct). Numbers have fallen in recent years and on conservation 'Red List'.

HABITAT Breeds on moorland, especially in steep-sided rocky valleys and around rocky outcrops, often where rough pastures nearby. On migration has a liking for horse paddocks in spring and stands of blackberry bushes in autumn.

autumn/ winter

VOICE Call *chack-chack-chack* – rather like Fieldfare. Song loud, 'wild' and rather simple, with repeated phrases like a Song Thrush.

CONFUSION SPECIES White-collared adults are distinctive (but beware partial albino Blackbirds), while autumn immatures can be separated from female and immature Blackbirds by pale fringes to wing feathers, faintly scaled underparts and 'chacking' call.

FIELDFARE · 25.5 CM · *Turdus pilaris*

grey rump
conspicuous
in flight

adult

in flight
shows white
wing-lining,
as Mistle
Thrush

adult

One of the classic winter birds, whether seen in flocks on pastures or playing fields, or flying over in loose, noisy parties.

DESCRIPTION Large, rather long-tailed thrush. Head grey, separated from grey back and rump by rich brown saddle, tail black, underparts white, heavily streaked darker, and breast ochre. Sexes similar. Juvenile very like adult. In flight underwings contrastingly white.

POPULATION Common passage migrant and winter visitor from N Europe and Scandinavia (Oct–Apr). Also rare breeder, with a few pairs in N Britain, although has nested as far S as Kent.

HABITAT Open fields, hedgerows, orchards and parks. Will sometimes visit gardens, especially during hard weather, to feed on fruit. Usually in flocks, sometimes with Redwings. Often feeds on ground, flying up into tall bare trees when disturbed.

autumn migrants
feast on hedgerow
fruits

VOICE A characteristic, loud *shack-shack, shakshakshak…*, usually given in flight.

CONFUSION SPECIES Other thrushes, but Fieldfare is distinctive when seen well.

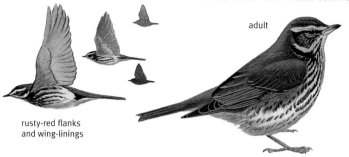

Turdus iliacus • 21 CM • **REDWING**

adult

rusty-red flanks and wing-linings

The smallest of the thrushes and a winter visitor to most of Britain, frequently found in mixed flocks with Fieldfares and other thrushes.

DESCRIPTION Upperparts dark brown, underparts off-white with numerous short soft streaks. Key features are bold whitish eyebrow and rusty-red flanks and underwing. Sexes similar, immature as adult.

POPULATION Common winter visitor from Iceland and Scandinavia, usually present late Sept–Apr. Also a rare breeding species, nesting in woodland in N Scotland, and exceptionally has bred as far S as Kent.

HABITAT Farmland, woodland, parks, and will visit gardens, especially in hard weather. Highly gregarious and usually found in flocks, feeding on the ground or on hedgerow fruit, but usually quite shy – the whole flock will fly off calling when disturbed.

VOICE A high-pitched but full *seehp*. This call can often be heard on autumn nights as migrants arrive from Europe for the winter.

CONFUSION SPECIES Distinguished from other thrushes by white eyebrow and reddish flanks. In flight shows reddish underwing; underwing is dull pale orange in Song Thrush and white in Mistle Thrush and Fieldfare.

bold whitish eyebrow

adult

adult

SONG THRUSH · 23 CM · *Turdus philomelos*

adult

juvenile

in flight shows pale orange wing-linings

cocks head to side to listen for worms

A familiar garden bird with a wonderful song, this is the commonest spotted thrush, although a significant decline puts it on the conservation 'Red List'.

DESCRIPTION Medium-sized thrush, spots on underparts very subtly V-shaped. Sexes similar. Juvenile spotted buff on upperparts.

POPULATION Common. Present all year, but a large proportion move to Ireland, France and Spain in winter, when upland areas largely deserted. Conversely, some European birds winter in Britain.

HABITAT Woods, hedgerows, parks – anywhere with trees or bushes. Smashes snails open on hard 'anvils', snails are particularly important during summer droughts and winter freezes when earthworms are hard to find.

VOICE A single *sip*. Song a series of strident whistled phrases, each repeated 2–4 times, e.g. *tui-tui-tui*, *titi-titi*, *didyu-didyu-didyu...*, and each bird may use over 100 different phrases. Higher-pitched than Blackbird, sweeter than Mistle Thrush, the repetition is characteristic.

CONFUSION SPECIES Smaller and shorter-tailed than Mistle Thrush, upperparts more uniformly brown, lacks white in tail-corners, and underparts often washed buff on breast and flanks. Lacks white eyebrow and reddish flanks of Redwing.

juvenile

adult

The largest thrush, has a habit of defending fruit-laden trees through the winter to ensure a supply of food, including large clumps of mistletoe, hence its name.

DESCRIPTION Large, long-tailed thrush with prominent whitish fringes to wing feathers and white underwing-coverts. Sexes similar. Juvenile much more variegated above, with many whitish spots.

POPULATION Fairly common but declining resident, but upland areas and N Scotland deserted in winter, when some young birds move to Ireland and France. Usually found in pairs, but in late summer and autumn frequently forms small flocks.

HABITAT Woods, mature hedgerows, parks, sometimes gardens; prefers areas with mixture of tall trees and grassland, but occasionally found on open moorland.

VOICE Loud, far-carrying song given in late winter and early spring, structure like Blackbird's song but phrases shorter, harsher and wilder. Often sings during inclement weather leading to old country name of 'Stormcock'. Call a distinctive dry rattle given in flight, recalling wooden football rattle.

CONFUSION SPECIES Larger than Song Thrush, white underwing and tail corners distinctive in flight (often flies high), on ground note very upright stance, greyer and more variegated upperparts and sparser and more rounded spots on underparts.

flight undulating with periods of flapping interspersed with periodic closure of wings

in flight note white wing-linings

BLACKBIRD · 25 CM · *Turdus merula*

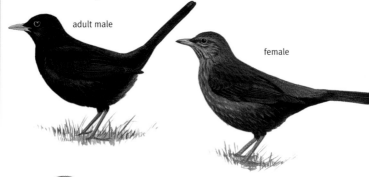

adult male

female

juvenile

1st-year male
has black bill

*One of the most familiar garden birds, with one of the
richest and loveliest songs. Often sings from rooftops
and TV aerials.*

DESCRIPTION Large thrush with long, full tail. Males
black, adults with narrow yellow eye-ring and orange-
yellow bill, immatures with dark bill. Females sooty-
brown with paler throat and brownish breast that is
vaguely streaked darker, bill varies from dark brown to
yellow. Juveniles cinnamon-brown with many small pale
spots and streaks; wings and tail darker.

POPULATION Common. Although many urban and
suburban birds are sedentary, there is some withdrawal
from higher ground in winter when some movement of N
breeders to Ireland and S breeders to France.
Conversely, large number of immigrants from Europe
move to Britain for the winter.

HABITAT Woods, hedgerows, parks and gardens.

VOICE Song slow, rich and melodic, without repetition.
In alarm or at dusk gives a loud *pik, pik, pik...*, which can
become an explosive rattle when flushed.

CONFUSION SPECIES Male could be confused with
Starling but lacks spots and has dark legs; when on the
ground long tail is often flirted upwards and often stops
dead for several seconds. Females and juveniles,
although diffusely spotted or streaked below, are much
darker than any other thrush except Ring Ouzel.

Locustella naevia • 13 CM • **GRASSHOPPER WARBLER**
Cettia cetti • 13.5 CM • **CETTI'S WARBLER**

Grasshopper Warbler – note long, full tail

Cetti's Warbler adult

Grasshopper Warbler adult

These secretive warblers have distinctive songs: Grasshopper Warbler has an insect-like song, hence its name; Cetti's Warbler's loud outbursts are unmistakable.

Cetti's Warbler

DESCRIPTION Grasshopper Warbler is brown above with diffuse streaking, has a plain 'face', and is whitish below with buffier flanks and dark streaks on undertail-coverts. Often walks on ground. Cetti's has chestnut-brown upperparts and grey 'face' and underparts, with paler throat and darker, browner rear flanks and belly. Often skulks low down inside bushes.

POPULATION Grasshopper Warbler is an uncommon summer visitor (May–Jul), has decreased recently and is on conservation 'Red List'. Cetti's Warbler population was around 2,350 pairs in 2009, mostly in SE coastal areas.

HABITAT Both occur in marshes and damp scrub, Cetti's especially in lake- and riverside bushes. Grasshopper Warbler is also found in young conifer plantations.

Grasshopper Warbler

VOICE Grasshopper Warbler has an insect-like, mechanical reeling, lasting up to a minute at a time. Cetti's Warbler gives a sudden, loud outburst: *che-che, pi, CHEWEE-CHEWEE-CHEWEE.*

CONFUSION SPECIES Grasshopper Warbler recalls Sedge Warbler but lacks supercilium (see also Dunnock). Cetti's is similar to Reed Warbler but fatter, darker and has grey 'face' and underparts. Nightingale is larger, with a redder tail and paler underparts.

Cetti's Warbler

REED WARBLER · 13 CM · *Acrocephalus scirpaceus*

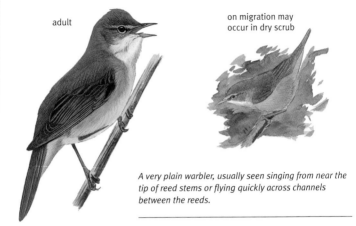

adult

on migration may occur in dry scrub

A very plain warbler, usually seen singing from near the tip of reed stems or flying quickly across channels between the reeds.

DESCRIPTION Slender warbler with rather long, pointed bill. Very plain, lacking any prominent field marks, although rump usually paler and more rusty-brown. Under-tail coverts long and full, and tip of tail graduated. Sexes similar. Juvenile as adult but overall rather warmer brown.

POPULATION Common summer visitor (Apr–Oct). Winters in tropical Africa. A frequent host of the Cuckoo and those that specialise in parasitising Reed Warblers have evolved eggs that closely match their host's.

HABITAT Strongly associated with reeds, whether in large reedbeds or bands of reeds fringing rivers, lakes and gravel pits. Migrants may, however, be found in drier scrub.

VOICE Call a harsh *churrr*. Song a rapid series of grating phrases, each repeated 2–3 times: *churr-churr-churr, chipit-chipit, che-che-che, chidi-chidi….*

CONFUSION SPECIES The only uniformly brown warbler commonly found in marshy places. Separated from Willow Warbler and Chiffchaff by larger size, longer, heavier bill and lack of any yellow in plumage; from Garden Warbler by bill shape and usually also habitat.

juvenile

adult

The commonest wetland warbler, occurring in a wide variety of marshy vegetation. Bold and often showy singing from bushtops or in a shorty song-flight.

DESCRIPTION Upperparts warm brown, diffusely streaked (can appear plain in a distance), with fairly bold off-white eyebrow contrasting with darker crown. Underparts white, washed buff on flanks and breast. Sexes similar. Juvenile as adult but usually has fine dark streaks on breast.

POPULATION Fairly common summer visitor (late Apr–Sept). Numbers have periodically crashed following droughts in African winter range.

HABITAT Reedbeds and other areas of tall marshy vegetation, especially where scattered small bushes provide songposts. Occasionally breeds in drier scrub.

VOICE Call a grating *trrrr* and hard *tak*. Song an 'angry' mass of harsh grating notes with occasional sweet whistled crescendos, faster than Reed Warbler and lacking the monotonous rhythm. Song sometimes given in flight.

CONFUSION SPECIES Streaked upperparts separate it from all other warblers except much scarcer Grasshopper Warbler, which is more boldly streaked, lacks a pale eyebrow and is seldom seen except when giving its prolonged buzzing song.

in flight shows warmer, more yellowish-brown rump

GARDEN WARBLER · 14 CM · *Sylvia borin*

adult

very nondescript, but note
stubby bill and mouse-
grey sides of neck

*Perhaps the most nondescript warbler and often very
skulking, but when seen well has a subtly distinctive
character and is, moreover, an excellent songster.*

in flight note
shorter tail
than Blackcap

DESCRIPTION Overall very plain, but note dark beady
eye in a 'gentle' open face, mouse-grey patch on sides of
neck, greyish legs and stubby blue-grey bill. Sexes
similar. Juvenile as adult.

POPULATION Fairly common summer visitor (late
Apr–Sept). Winters in tropical Africa.

HABITAT Prefers areas of dense scrub with some taller
trees or large bushes, including woodland rides and
edges, small copses, hawthorn and blackthorn thickets,
dense gorse on heathland and overgrown hedges.
Migrants may turn up in any bushy area.

VOICE Call an undistinguished and nervous *che*. Song
comprises fast, rambling phrases, 3–8 seconds long,
some notes full and fruity (recalling Blackbird in tone),
often given for prolonged periods with brief pauses
between each phrase. Overall rather unvaried, always
lacking the high sweet notes of Blackcap. Usually sings
from dense cover and more often heard than seen.

CONFUSION SPECIES Separated from Reed Warbler,
which is also rather plain and brown, by short, stubby
bill. Willow Warbler and Chiffchaff are smaller and
usually more olive-green, show yellow in plumage and
have rather more slender bill.

Sylvia atricapilla • 13 CM • **BLACKCAP**

female

male

A distinctive warbler, its dark cap being unique, and one of our very best songsters. An increasingly common visitor to garden bird feeders in the late winter.

DESCRIPTION Upperparts grey, cap black in males and warm brown in females, placed at a jaunty angle on head and not covering eye. Bill and legs grey.

POPULATION Fairly common. Primarily a summer visitor (Apr–Oct), most of population winters in Mediterranean basin. Conversely, small numbers of European breeders move to Britain and Ireland in winter.

HABITAT Woodland, parks and mature gardens with well-grown trees. In winter a regular visitor to garden bird feeders, especially from Christmas onwards.

VOICE Call a hard *tak*. Song initially a rambling chattering (rather like Garden Warbler), but each phrase typically ends with a crescendo of high, beautifully sweet, whistled notes. Sometimes gives several phrases without the sweet terminal notes and then very easy to confuse with Garden Warbler, but always eventually produces characteristic pure whistles. There are usually also significant gaps between each song phrase, unlike the near-continuous warbling of Garden Warbler.

CONFUSION SPECIES Separated from all other warblers by dark cap, and from Marsh and Willow Tits (which also have a black cap), by lack of black bib and more graceful shape, and also by voice and behaviour.

shares hard *tak* call with Lesser Whitethroat, but has no white in tail

WHITETHROAT · 14 CM · *Sylvia communis*

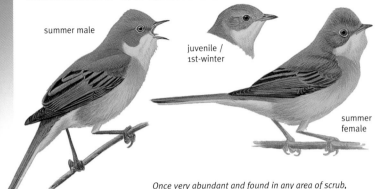

summer male

juvenile / 1st-winter

summer female

shows white outer tail feathers in short song flight

Once very abundant and found in any area of scrub, droughts in the Sahel region of Africa have caused periodic population crashes.

DESCRIPTION Bulky, long-tailed warbler with broad rusty-brown fringes to wing feathers, pale base to bill and yellow-brown legs. Male in spring has grey head with white eye-ring and reddish eye, pinkish breast and contrasting white throat (often puffed out). Female and immatures duller, with less conspicuous eye-ring, head and breast washed brown and less contrasting off-white throat.

POPULATION Fairly common summer visitor (mid Apr–mid Oct); winters in tropical Africa. Numbers have never fully recovered from drought-provoked crashes, the most dramatic was a 75% drop in numbers in 1969.

HABITAT Scrub – overgrown hedges, woodland edges, young coppice, bushy areas on dunes, heaths and downland, overgrown railway lines, etc. Migrants can occur in any scrubby area.

VOICE Call a distinctive *vedd-vedd-vedd*, in alarm a harsh *charrr*. Song a short, scratchy warble, varying little from bird to bird. Usually sings from an exposed perched or in short, showy song flight.

CONFUSION SPECIES Larger and longer-tailed than Lesser Whitethroat (which also has white outer tail feathers), with rusty in wing and pale legs and eyes.

1st-winter

adult

A small, neat warbler that keeps to cover but gives itself away with its song, a quiet warble followed by a loud rattle, audible for quite a distance.

DESCRIPTION Plain grey-brown above with greyer head, grey tail and white outer tail feathers. Ear-coverts usually darker than remainder of head, forming a dark mask, but this is sometimes not obvious. Narrow pale eye-ring, eye and bill dark. Sexes similar, immature very like adult.

POPULATION Fairly common summer visitor (May–Sept). Winters in Ethiopian Highlands of NE Africa.

HABITAT Breeds in dense thickets – overgrown hedgerows, abandoned railway lines, downland scrub, around gravel pits, even industrial sites. Migrants can occur in any area of scrub.

VOICE Call a hard *tak*. Song a quiet, undistinguished warble followed by a loud, very characteristic machine gun-like rattle:... *cha-cha-cha-cha-cha*. Usually sings from deep cover.

CONFUSION SPECIES Easily confused with Whitethroat, but note smaller size, overall neater and more compact appearance, lack of rusty-brown in wings, dark eyes and slate-grey legs. Distinguished from most other warblers by white outer tail feathers.

looks pallid below
on short flight
between cover

DARTFORD WARBLER · 12.5 CM · *Sylvia undata*

female

long tail often
cocked upwards

song flight

male

Named after Dartford in Kent and one of the few warblers resident in Britain all year round. It often buries itself in cover, but may be seen perched on bush tops.

DESCRIPTION Upperparts grey, underparts dark wine-red with small white spots on throat. Eye reddish, encircled by narrow, bright red eyering. Females duller, with pale grey throat. Immatures duller still, dusky grey with paler throat, dull red eye ring and brown eye.

POPULATION Locally fairly common, but vulnerable to harsh winter weather; British population fell to just 10 pairs following exceptionally hard winter of 1962/63. Numbers then peaked at around 3,200 pairs in 2006, but have fallen again significantly following heavy snowfalls in recent winters.

HABITAT Heathland with heather and gorse.

VOICE Song a fast, scratchy and rather tuneless phrase, varying little in pitch and with just a few sweet notes; it is rather like Common Whitethroat's, but less structured. Call a harsh *chairr, chairr*, again like Common Whitethroat (or distant, muffled Jay).

CONFUSION SPECIES Adults distinctive when seen well. Long-tailed Tit often found on heathland, but has even longer tail and very different plumage. Common Whitethroat also found in similar habitats, but shorter-tailed and much paler overall.

Phylloscopus sibilatrix • 12 CM • **WOOD WARBLER**

sings with whole body quivering

adult

Commands attention with a dramatic and loud trilling song, but seldom seen away from its woodland breeding grounds.

DESCRIPTION Upperparts moss-green with conspicuous paler edges to wing feathers and prominent yellow eyebrow. Throat and breast primrose-yellow, remainder of underparts silky-white. Sexes similar, immature as adult.

POPULATION Summer visitor (early Apr–Aug); winters in tropical Africa. Fairly common in W Britain, but has slowly declined in SE and East Anglia. Conversely, slowly increasing in Ireland, although still rare there. Infrequently seen on migration.

HABITAT Breeds in woodland with a high, closed canopy and sparse understorey, often on slopes, favouring stands of Beech and Sessile Oak.

VOICE Call a melancholy *piu*, recalling Bullfinch. Song has two variants. Commonest is stuttering series of sharp, loud *zip* notes that accelerates into a very fast shivering trill, its whole body shaking; often sings in short gliding song flight. Less frequently gives a series of pure, whistled *piu* notes.

CONFUSION SPECIES Scarcest of the three breeding leaf warblers, separated from Willow Warbler and Chiffchaff by cleaner green upperparts and the sharp contrast between white underparts and yellow breast.

WILLOW WARBLER · 11 CM · *Phylloscopus trochilus*

juvenile

spring adult

The commonest of the leaf warblers, a group of three small, generally greenish warblers that are always on the move, flitting in and out of view amongst the leaves.

DESCRIPTION Upperparts olive-green, underparts pale, often with subtle yellow tones. Pale eyebrow, though not prominent, often more obvious than pale eye-ring. Sexes similar. Juvenile as adult but often a little brighter and yellower. Frequently flicks its wings.

POPULATION The commonest summer visitor to the British Isles, present early Apr–early Oct although there has been a recent decline in Souther England. Winters in tropical Africa.

HABITAT Woodland, favouring bushier areas along edges, rides and in clearings, also scrub, young conifer plantations and occasionally scrubby rural gardens. Less closely associated with tall trees than Chiffchaff.

VOICE Call *huweet*, very like Chiffchaff. Song a delightful series of soft whistles that descends the scale; rhythm rather like a Chaffinch's song (although lacking a terminal flourish), but much sweeter and more silvery in tone.

CONFUSION SPECIES Very similar to Chiffchaff, silent birds are best separated by paler and browner legs and obvious pale base to bill. Unlike Chiffchaff, very rarely recorded in winter.

Phylloscopus collybita · 10.5 CM · **CHIFFCHAFF**

juvenile
plumage
brighter and
fresher

spring adult

This leaf warbler is one of the few British birds to call its name, and its plaintive song delivered from still-leafless trees is one of the first real signs of spring.

DESCRIPTION Upperparts olive-green, underparts pale but rather drab. A pale eye-ring, though not prominent, is often more obvious than the pale eyebrow. Sexes similar. Juvenile as adult, although a little brighter. Often flicks its wings and tail.

POPULATION Common summer visitor (late Mar–Oct). Most winter around the Mediterranean or in Africa S of Sahara, but a small minority overwinter in Britain, especially in milder S and W, and these are supplemented by small numbers from the Continent.

HABITAT Breeds in open woodland and scrub with at least some mature trees. Migrants can be found in wide range of scrubby habitats, including quite small gardens. In winter favours scrubby places near water, also reedbeds and sewage tanks.

VOICE Call *huweet*, very like Willow Warbler. Song a slightly halting repetition of two notes in an irregular pattern: *chiff-chaff chiff-chaff chiff-chiff, chiff-chaff...*, often introduced by one or two quieter grating notes.

CONFUSION SPECIES Very like Willow Warbler, and when not singing best separated by dark legs and predominantly dark bill (although some Willow Warblers also have dark legs).

SPOTTED FLYCATCHER · 14.5 CM · *Muscicapa striata*

adult

juvenile

Characteristically perches rather upright, flicks its wings, and sallies out to catch insects before returning to the same or a new perch.

sallies out to catch
flying insects

DESCRIPTION Drab, sparrow-sized bird with a fine but broad-based bill, long wings and a long tail, giving a slender profile. Note narrow pale fringes to wing feathers, plain face with dark, beady eye, black bill and legs, fine streaks on forecrown and diffuse streaks on breast. Sexes similar. Juvenile as adult but spotted buff on upperparts.

POPULATION Summer visitor (May–Sept); winters in S Africa. Uncommon, has been declining steadily since the 1960s and now on the conservation 'Red List'.

HABITAT Woodland rides and clearings, and parks and gardens with mature trees. Frequently sits on exposed branches or posts. Often nests in tangles of honeysuckle and other climbers in gardens.

VOICE Call a high, almost hissing *tseeh*, in alarm a more noticeable *see-chik* or *see-chik-chik*. Song unobtrusive, thin and squeaky.

CONFUSION SPECIES Distinctive if seen well, the flycatching behaviour is good clue. Female and immature Pied Flycatcher are always much neater, with a prominent white wing-patch.

Ficedula hypoleuca • 13 CM • **PIED FLYCATCHER**

female/autumn
male/1st-winter

summer male

A very smart flycatcher, most easily found in spring in the Sessile Oak woods of the uplands of N and W Britain but also widespread on migration on E and S coasts.

DESCRIPTION Sparrow-sized, neat and compact. Breeding male black and white. Female, autumn male and immature clean brown above and off-white below with faint buff wash on breast. Note whitish patch in wing and narrow white outer tail feathers. Juveniles spotted above. Sallies out after insects, usually returning to a different perch, and often cocks tail and flicks one wing.

POPULATION Fairly common summer visitor to the west of the UK (May–Sept), winters in W Africa. Conspicuous on breeding grounds when in song, becoming elusive in midsummer.

juvenile

HABITAT Breeds in deciduous woodland, especially upland oak woods; takes readily to nest boxes. On migration in Aug–Sept could turn up anywhere with trees or bushes, but commonest on E and S coasts.

VOICE Call an abrupt *pik*. Song a slow series of simple notes, going up and down the scale.

CONFUSION SPECIES Flycatching habits distinctive; females and autumn birds separated from Spotted Flycatcher by white in wings and neat unstreaked appearance. Vaguely similar female Chaffinch may make flycatching sallies at times but note bill shape.

FIRECREST · 9 cm · *Regulus ignicapilla*

adult male

adult

adult female
lacks orange
on crown

juvenile

This jewel is one of Britain's most attractive birds. It first bred in 1962 and has slowly built up its numbers - breeding birds are best located by song.

DESCRIPTION Upperparts greenish, with bold black-and-white bars across wings and gorgeous flame coloured shoulder. Head marked with broad white stripe above eye, and crown usually shows some orange in males but clean yellow in females. Underparts off-white.

POPULATION Around 500 pairs breed in Britain, with main concentrations in Norfolk, Suffolk, the Chilterns, Sussex and, especially, Hampshire. More widespread on passage (Mar–May and Oct–Nov), when it can turn up almost anywhere, but commonest on the S coast. Small numbers winter in the S and in Wales.

HABITAT Breeds in mixed and coniferous woodland, favouring stands of mature Norway Spruce. On migration and in winter found in woodland and scrub, usually on or near the coast.

VOICE Song a high-pitched, rhythmic *tsi-zi-zi-zi-zi-zi-zi-zi-ZI-ZI-ZI*, increasing in volume towards the end. It is very like Goldcrest, but lacks its slightly halting, rolling, quality and, more obviously, lacks a terminal flourish. Call *tzi-tzi*, subtly harder and fuller than Goldcrest.

CONFUSION SPECIES Similar to Goldcrest, but easily separated by white line above eye and overall brighter and greener upperparts.

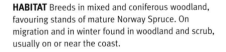

adult

tiny but capable
of crossing the
North Sea

*The smallest British bird, weighing around 6.5 g. Usually
attracts attention with its very thin, high-pitched calls
(so high-pitched some older people cannot hear them).*

DESCRIPTION Tiny. Sides of crown black, framing
yellow central stripe. Dark beady eye, surrounded by
a broad but diffuse whitish eye-ring, gives a rather
startled expression. Sexes similar, but males have
orange base to yellow crown feathers, exposed when
excited. Juvenile very plain-headed with pale bill. Very
active, flicking its wings and often hovering. In late
summer and winter found in mixed flocks together
with tits.

when excited
male reveals
orange bases to
crown feathers

POPULATION Common resident, numbers are
supplemented in winter by Continental birds.

HABITAT Closely tied to conifers, breeding in coniferous
and mixed woodland. More widespread in winter and
regular in gardens, especially when conifers are present,
but seldom visits feeders. Usually forages high in
canopy.

juvenile
lacks yellow
on head

VOICE Call *si-si-si*. Song a rhythmic repetition of the call
ending with a terminal flourish: *zee-zee-zee-zee-zee-zee-
ziddly-zee ziddle-u.*

CONFUSION SPECIES Head pattern distinguishes it
from all warblers. The closely related Firecrest is
brighter, with a bold white eyebrow.

BEARDED TIT · 12.5 CM · *Panurus biarmicus*

male

female

juvenile
female

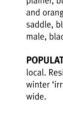

juvenile
male

Not a tit (indeed, not even closely related to other tits), this charismatic reedbed specialist is often hard to see, but attracts attention with its distinctive 'pinging' calls.

DESCRIPTION Adult male has blue-grey head and striking black moustache (not a 'beard'); female much plainer, but note extensive white in wing, yellowish bill and orange eye. Juveniles more colourful, with black saddle, black and white on sides of tail and in juvenile male, blackish mask between eye and bill.

POPULATION Rather uncommon (c. 400 pairs) and very local. Resident, but subject to occasional autumn and winter 'irruptions' when small parties disperse far and wide.

HABITAT Breeds in large reedbeds, eating insects in summer and seeds in winter. Winter wanderers may be found in quite small stands of reeds and other tall marshy vegetation.

VOICE A distinctive explosive, metallic *teu*, given in an irregular pattern; may be taken up as a chorus by several birds.

CONFUSION SPECIES Males unique, females and juveniles drabber but note reedbed habitat, long tail, overall pale warm brown coloration and bold wing markings. Reed and Sedge Warblers are much plainer, shorter-tailed and less acrobatic.

Aegithalos caudatus • 14 CM • **LONG-TAILED TIT**

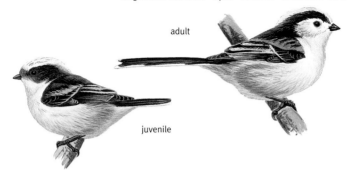

adult

juvenile

One of Britain's smallest birds, weighing 7–9 g. Builds a ball-shaped nest, camouflaged with up to 3,000 flakes of lichen and lined with around 1,500 small feathers.

DESCRIPTION A tiny 'ball of fluff' with a very long, narrow tail (*c.* 60% of total length). Sexes similar. Juvenile shorter-tailed and overall much duller and sootier.

POPULATION Generally fairly common resident. Highly gregarious, usually in flocks moving 'follow-my-leader' through the vegetation. Even breeding pairs are often aided by extra 'helpers', usually siblings of the male. Its small size makes it vulnerable to cold winters (after which numbers may fall) and members of a flock roost huddled together to conserve heat.

gregarious and almost always seen in pairs or flocks

HABITAT Mature scrub – woodland edges and glades, heaths, mature hedgerows; less often suburban gardens, parks and cemeteries. Food almost entirely insects and spiders, but an increasingly frequent visitor to bird feeders.

VOICE Very vocal, flocks keep in contact with a thin but penetrating, high-pitched *see-see-see* and a lower-pitched, mellower clipped *tup*; in excitement gives a louder and harder slurred, rattling *tsirrrup*.

CONFUSION SPECIES
Small size, long tail, and black, white and pinkish coloration unique.

CRESTED TIT · 11.5 CM · *Lophophanes cristatus*

adult

adult

adult

Britain's only crested tit, and also the most localised of the tits, being confined to pinewoods in the highlands of Scotland. Sometimes visits bird feeders.

DESCRIPTION Upperparts brown, underparts off-white, washed buff on flanks. Head pattern distinctive, with black line through eye and around ear-coverts, black bib and pointed, black-and-white spangled crest. Sexes similar, and juveniles much like adults.

POPULATION Fairly common within limited range, with around 2,400 pairs. In winter months may join with other species of tit to form roving flocks, but can be unobtrusive and best located by listing for purring call.

HABITAT Pinewoods, both native 'Caledonian Pinewoods' (relicts of the once more extensive primeval woodland that covered most of Scotland) and plantations of Scots Pine.

VOICE Most characteristic call, given by foraging birds, a soft, slightly stuttered, purring trill, usually combined with more typically tit-like thin *si-si* call, thus *si-si-si-trrrrrrrrrrrrr*.

CONFUSION SPECIES The only small bird in Britain to sport a crest. Coal Tits occasionally have a hint of a crest, but nothing more than a ruffling of the nape.

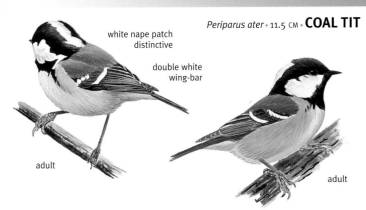

Periparus ater • 11.5 CM • **COAL TIT**

white nape patch
distinctive

double white
wing-bar

adult

adult

*The smallest tit, making up for its inability to lay down
large stores of fat by hiding numerous pieces of food,
such hoards being used when food is scarce.*

DESCRIPTION Typical perky tit, but overall rather dull.
Note white cheeks, nape-patch and double wing-bar.
Sexes similar. Juvenile has nape, cheeks and underparts
washed dull yellow.

POPULATION Common resident, but wanders a little in
winter.

HABITAT Strongly linked to coniferous woodland, from
natural Scots Pine in the Highlands to plantations of
exotics and even mature specimen trees in gardens; will
forage in nearby deciduous trees, and breeds in oak and
birch in N and W. Usually feeds in tree tops, but will
forage on ground; a frequent visitor to bird feeders.
Generally nests in lower counties than Blue or Great Tits,
sometimes holes amongst roots or in the ground, and
occasionally uses nestboxes.

VOICE Song a ringing *pitch-u, pitch-u, pitch-u...*, easily
confused with Great Tit but rather faster and shriller. Has
a variety of high-pitched calls, some of which recall
Goldcrest.

CONFUSION SPECIES Drab coloration and white nape
unique amongst tits. Strong yellow tones of juveniles
invite confusion with young Great and Blue Tits, but pale
nape already present.

extensive
black bib

juvenile
washed
yellowish

GREAT TIT · 14 CM · *Parus major*

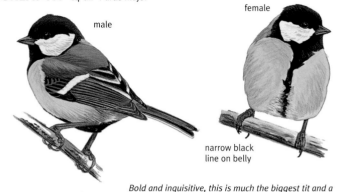

male

female

narrow black line on belly

white outer tail feathers

juvenile

Bold and inquisitive, this is much the biggest tit and a regular visitor to bird feeders. This species is almost as common and familiar as the ubiquitous Blue Tit.

DESCRIPTION Cap black, cheeks white, upperparts greenish with a single whitish wing-bar, underparts yellow with bold black central stripe extending down from black bib. Sexes rather similar, but male has broader black stripe on belly. Juvenile duller, with black replaced by sooty grey and cheeks yellowish. Usually travels in pairs, but will join mixed flocks of tits, etc.

POPULATION Common resident.

HABITAT Essentially deciduous woodland, but adaptable and found in coniferous woodland, farmland and suburban parks and gardens. Breeds anywhere with suitable nest holes, usually in trees but is innovative and takes readily to nest boxes. Often feeds on ground and partial to beech mast in the winter.

VOICE Song loud and ringing, a repetitive *tea-cher, tea-cher...* or similar two-note phrase, given Jan–May. Other calls highly variable and often confusing, 'if you don't recognise it, it's probably a Great Tit'.

CONFUSION SPECIES Rather bigger than Blue Tit, with a solid black cap, white outer tail feathers and broad black stripe down breast and belly.

female

male

A familiar and charismatic visitor to bird feeders, this is the commonest tit and a national favourite. Takes readily to nest boxes.

DESCRIPTION Cap ultramarine blue, wings and tail blue, underparts yellow. Sexes very similar. Juveniles rather drabber, with yellowish face and greenish cap.

juvenile

POPULATION Common resident.

HABITAT A bird of deciduous woodland, especially oak; has adapted well to a man-made environments and found in well-grown hedgerows, parks and gardens. Rather arboreal and seldom visits the ground. Lays up to 16 eggs, and laying is timed to exploit spring flush of woodland caterpillars when eggs hatch.

no white in tail

VOICE Song a clear, silvery, high-pitched trill lasting two seconds, usually introduced by 1–4 high-pitched notes: *pee-pee-ti sihihihihihi, pee-pee-ti sihihihihihi....* Other calls very varied.

CONFUSION SPECIES Separated from Great Tit by blue cap, white eyebrow and cheeks separated by irregular black line through eye, all-blue tail and yellow breast, with black on underparts confined to faint narrow line on belly.

MARSH TIT · 11.5 CM · *Poecile palustris*
WILLOW TIT · 11.5 CM · *Poecile montanus*

Willow Tit
adult

note pale panel
on closed wing

Marsh Tit
adult

wings
uniform

These two tits are very hard to separate on plumage, and even their voices confuse many experienced birdwatchers.

DESCRIPTION Both are brown above with a black cap and bib. In fresh plumage in autumn and winter Willow Tit has a paler and greyer panel on the closed wing (formed by pale fringes) lacking in Marsh Tit. Other differences are too subtle and subjective to be of use.

POPULATION Marsh Tit is fairly common, Willow Tit uncommon, but both species are on the conservation 'Red List' due to recent sharp declines.

HABITAT Marsh Tit favours deciduous woodland, and is a regular visitor to bird feeders in nearby gardens. Willow Tit breeds in damp woodland and disperses in winter to hedgerows, but seldom visits feeders.

VOICE Marsh Tit is noisy with a variety of calls including a nasal *zee-zee-zee*, but will eventually gives its characteristic sneezing *pitch-u*. Willow Tit is much quieter, its calls are usually limited to a thin *sit* or *si-si* and very nasal *dzee-dzee-dzee*. Marsh Tit's song is loud and full but very variable, usually a rattled repetition of 8–20 simple notes. Willow Tit's song is a pure, ringing, whistled *piu-piu-piu...* (occasionally closely matched by Marsh Tit).

CONFUSION SPECIES Separated from Coal Tit by lack of wing-bars or white nape-patch.

Marsh Tit

Willow Tit

shows white
in tail corners
in flight

adult

Distinctive. Forages on trunks of large trees and uniquely able to climb downwards head-first. Unlike woodpeckers and Treecreeper does not use the tail as a prop.

DESCRIPTION Squat and short-tailed with long, dagger-like bill. Sexes very similar, but male has darker and more contrasting brick-red flanks and vent. Juvenile like adult female. Often attracts attention by loudly hammering at nuts and seeds wedged into crevices.

POPULATION Resident, highly territorial, and found in pairs throughout the year. Range has slowly spread N to Scotland in recent decades. Fairly common, especially in regions with much woodland.

HABITAT Deciduous woodland, parks and sometimes large gardens, with plenty of mature trees to provide suitable nest holes. Uses mud to plaster-up the entrance to its nest until it is a tight fit, helping to keep out larger competitors. Will sometimes forage on ground and visits bird feeders.

VOICE Very vocal, giving loud piping and trilling calls from tree tops, including a repeated, loud, whistled *dwip-dwip*; song a repetition of similar whistles, varying from slow and deliberate to fast and trilling. Contact call less conspicuous, a high thin *tsit*.

CONFUSION SPECIES Blue-grey upperparts, bold black eye-stripe and tree climbing habits unique.

gives thin *tsit*
calls in flight

TREECREEPER · 12.5 CM · *Certhia familiaris*

adult

circles each tree
in turn in search
of food

*An inconspicuous, mouse-like bird that shuffles jerkily
up and along the trunk and larger branches of trees in
search of insects and spiders.*

DESCRIPTION Mottled upperparts provide good
camouflage against tree trunks, white underparts often
hidden. Bill fine and decurved for probing bark fissures,
tail stiffened, like a woodpecker's, to provide a prop
when climbing. Sexes similar. Juvenile as adult.

POPULATION Fairly common resident.

HABITAT Woodland, both deciduous and coniferous, as
well as mature conifer plantations. Will disperse to
parks, gardens and farmland with suitable old trees.
Never visits bird feeders. Often found singly, but will join
tit flocks.

VOICE Calls include a thin but penetrating and high-
pitched, almost hissing *tsree*, often repeated (this call is
harsher than otherwise similar call of Goldcrest). Song a
strident descending warble with final hurried flourish,
lasts about three seconds, overall rhythm recalls
Chaffinch or Willow Warbler.

CONFUSION SPECIES Mottled brown upperparts, fine
decurved bill and tree-climbing habits unique.

roosts
fluffed up
in a bark
crevice

Pyrrhocorax pyrrhocorax • 40 CM • **CHOUGH**

bill and legs red

This charismatic crow is confined to the W coasts of Britain and Ireland, but recently and unexpectedly returned to England when a pair bred in Cornwall in 2002.

DESCRIPTION Red bill and legs can be hard to see at a distance. In flight note broad, well-fingered wingtips, can be very acrobatic. Sexes similar. Juvenile has brownish-yellow bill.

POPULATION Scarce. The current British population is around 450 pairs, with a further 800 in Ireland. Gregarious, often found in flocks.

HABITAT Breeds on coastal cliffs or in large inland quarries, but will sometimes nest in old buildings or even abandoned mineshafts. Requires old, unimproved but closely-grazed short-grass pastures for feeding, and this combination of specialist nesting and feeding requirements explains its rarity. Very sedentary and seldom seen away from breeding areas.

VOICE Noisy, the characteristic flight call is a nasal *chaaw*, sometimes squeaky, sometimes groaning.

CONFUSION SPECIES Red legs and bill very distinctive, as is flight silhouette, but Jackdaws can be equally acrobatic and can have confusingly similar calls.

wing-tips show prominent 'fingers'

MAGPIE · 45 CM · *Pica pica*

adult

'One for sorrow, two for joy...', the Magpie is so familiar that it has entered folklore. Intelligent and adaptable, it does well in urban environments.

places its untidy domed nest in thick hedges

DESCRIPTION Sexes similar. Juveniles duller, with a shorter tail.

POPULATION Common, especially in towns and cities, but absent in NW Scotland. Can be scarce in rural areas if there are vigilant gamekeepers. Flocks in winter and early spring, and sometimes as many as 100 birds can be seen together. Once persecuted for its impact on game birds, it is often blamed for having a negative impact on songbird numbers, but there is little evidence that Magpies are responsible for declines in populations of other birds. Strictly resident, Magpies do not move far from where they were hatched.

HABITAT Very varied, but prefers relatively open country, with scattered trees and bushes to nest in and usually absent from treeless moorland and dense woodland. Often seen scavenging road-kills.

VOICE A characteristic harsh, rattling *cha-cha-cha-cha-cha*.

CONFUSION SPECIES None.

Garrulus glandarius • 34 CM • **JAY**

adult

conspicuous
white rump

*Dusty pink with bold black and white markings and a
beautiful blue shoulder, this colourful crow is one of our
most exotic looking birds.*

DESCRIPTION Often seen in flight, slipping from tree to
tree or flying longer distances from wood to wood.
Wings broad and rounded, flight floppy. Black tail and
white on rump and wings conspicuous, blue on wings
much less so. Sexes similar. Juveniles as adult.

POPULATION Common resident, except in N Scotland.
Numbers have slowly recovered since the 1940s from
heavy persecution in the 19th and early 20th centuries.
Usually seen singly or in pairs, often feeds on ground,
but can be shy.

HABITAT Deciduous woodland, also hedgerows, parks
and gardens with mature trees. Jays have a special
relationship with oak trees and acorns form a large
proportion of their diet. They are the main agents for the
dispersal of oaks and in autumn hide around 2,000
acorns for later consumption; inevitably, they forget to
retrieve a few, which go on to produce new trees.

VOICE A loud, harsh, screeching *shreeek, shree-shreeek*.

CONFUSION SPECIES None. Often reported as
something 'strange' or 'exotic' by non-birdwatchers.

can raise its short
crest when excited

JACKDAW · 33.5 CM · *Corvus monedula*

adult

1st-winter has duller eyes

The smallest of the crows, Jackdaws have a reputation for inquisitiveness and opportunism, and have adapted well to urban and suburban environments.

often found around towns and villages

DESCRIPTION Greyish-black with a pale grey cowl covering sides of neck and nape. Sexes and ages very similar. Flight fast, flappy and agile, note short bill and short, broad neck.

POPULATION Common, although scarcer in upland areas. Present all year, but some movement towards the milder climes of W Britain and Ireland during winter, with European birds simultaneously moving into England. Mates for life and highly sociable, almost always seen in pairs or flocks, often mixed with Rooks. Forms large communal winter roosts, gathering in noisy acrobatic flocks towards dusk.

HABITAT Very varied, but requires combination of grassland or arable for feeding (preferring areas of livestock or mixed farming) and suitable nest sites. Most nests are placed in holes, from cavities in sea cliffs and old quarries, to rabbit burrows, tree holes and old buildings. Often nests in loose colonies.

VOICE A distinctive, explosive *jack* (hence its name) and buzzing *kaarrr*.

CONFUSION SPECIES Smallest crow. At a distance appears blackish, but seen well grey cowl and pale eye distinctive. Hooded Crow has an entirely grey body.

Corvus corax • 64 CM • **RAVEN**

juvenile

plumage
duller and
less glossy

The largest crow, heavy persecution in the past confined Ravens to the remote N and W, but they are very slowly making a comeback.

DESCRIPTION All black with heavy, deep-based bill and thick, sometimes shaggy throat feathering (a 'beard'). Sexes similar. Juvenile as adult but duller, more sooty-brown. Often soars or engages in aerial acrobatics, especially half-rolls.

POPULATION Uncommon resident but numbers increasing and gradually spreading east across Britain. Mates for life and usually seen in pairs or small flocks. Frequently shy and wary.

HABITAT Very adaptable and able to thrive in many habitats but mostly confined to moorland and sea cliffs and areas where livestock farming provides suitable carrion (and thus pushed out by blanket forestry or switch to arable farming). Breeds very early, building large stick nests on cliffs or trees.

VOICE Very varied. Most characteristic is a deep, powerful *pruk*, but many other calls, some rather odd.

CONFUSION SPECIES Much larger than other crows (bigger even than a Buzzard). On ground note big bill and shaggy beard. In flight relatively long, narrow wings, long, diamond-shaped tail, well-projecting head and bill distinctive, but without clues as to size, can recall Rook.

note
diamond-
shaped
tail and
prominent
head/bill

highly acrobatic

CARRION CROW · 46 CM · *Corvus corone*
HOODED CROW · 46 CM · *Corvus cornix*

Carrion Crow

Hooded Crow

Carrion Crow

Hooded Crow

Carrion Crow

Hooded Crow

Until recently this closely related pair were treated as a single species. They are easily separated on range and plumage.

DESCRIPTION Carrion Crow is all-black, whilst Hooded Crow has a grey body with black hood, wings and tail. In a belt from the Clyde to Caithness in Scotland their ranges meet and they hybridise commonly; the hybrids show an intermediate pattern.

POPULATION Both are common residents, usually seen singly or in pairs, sometimes small parties, and often roost communally in the winter, sometimes with Rooks and Jackdaws. Hooded Crow was a winter visitor to E England from Scandinavia but is now rare.

HABITAT Anywhere from city centres to mountain tops. Nest singly, usually in trees.

VOICE Abrupt, somewhat menacing *caw, caw-caw, caw-caw-caw....*

CONFUSION SPECIES See Rook and Jackdaw. Hard to separate Carrion Crow from juvenile Rook, note deeper, blunter bill. Compared to Raven, smaller with a finer bill.

Corvus frugilegus • 45 CM • **ROOK**

immature

adult

One of the characteristic sights and sounds of the countryside, especially the noisy spring antics of a tree top rookery.

DESCRIPTION Adults have greyish face produced by pale skin around base of bill and a distinctive profile with vertical forehead and high peaked crown. Sexes similar. Juvenile lacks pale face, which develops in its second year.

POPULATION Common resident, but avoids busy urban and suburban areas. Very sociable, usually feeding in flocks and in winter forming large communal roosts, often with Jackdaws.

HABITAT Farmland, favouring areas of mixed farmland. Often seen foraging on verges of busy roads. Breeds colonially, siting its nest in tall trees, such rookeries often located in villages or around farms.

VOICE A drawn-out, crowing *caaw, caaw-caaw....* The cacophony at a rookery can be deafening.

CONFUSION SPECIES Closest to Carrion Crow, but pale face of adult diagnostic. Juvenile very similar to Carrion Crow, but usually seen with adults and note deeper, more drawling call. In flight Rook very similar to Carrion Crow, but slightly more diamond-shaped tail and more obvious head/bill can suggest Raven.

tail subtly diamond-shaped

STARLING · 21.5 cm · *Sturnus vulgaris*

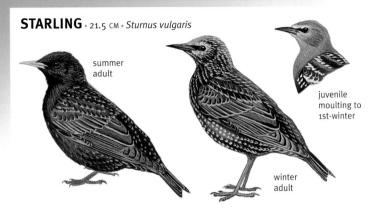

summer adult

winter adult

juvenile moulting to 1st-winter

This once abundant bird is on the conservation 'Red List' due to a rapid decline in recent years. It is still a familiar species throughout the UK, especially in winter.

juvenile

DESCRIPTION Numerous pale feather tips give spangled appearance in fresh autumn plumage, these wear away over winter and by spring plumage more uniformly blackish with patches of green and purple iridescence. Bill yellow in summer, dark brown in winter. Sexes similar, but in spring base of bill subtly blue in male, faintly pink in female. Juvenile distinct, the colour of strong tea; in late summer moulting juveniles are a ragged mixture of brown and black.

POPULATION Still fairly common, but breeding population has declined by over 50% in the last 25 years. Resident birds augmented by many winter visitors from Europe, and in winter Starlings roost communally, sometimes in vast flocks; the pre-roost gatherings produce a great spectacle.

HABITAT Ubiquitous, avoiding only remote moorland and mountains.

VOICE Song, given from high perches in trees or houses, is a musical whistled cacophony with many imitations of other birds as well as human artifacts.

CONFUSION SPECIES In worn summer plumage can be uniformly black, but smaller and shorter-tailed than Blackbird, has reddish legs, and walks rather than hops.

female / juvenile

summer male

Once common, the cheeky sparrow has undergone a marked and mysterious decline and is now on the 'Red List' of endangered British birds.

head pattern more subdued

winter male

DESCRIPTION Upperparts dull brown streaked blackish, underparts dirty pale grey. Male has black mask between eye and bill, large ragged black bib and chestnut sides to grey crown. Female and immature much drabber, lacking black face and bib but often showing narrow pale eyebrow.

POPULATION Resident; still common in some areas, but has declined by over 64% in recent years. The causes of the sparrow's demise are poorly understood, but probably include shortages of insect food for nestlings in summer and loss of breeding and roosting sites.

male

HABITAT Closely associated with man, the majority breed in villages and urban areas. Commonly visits bird roosts communally, frequently in ivy thickets.

VOICE Varied chirps and chatters.

CONFUSION SPECIES Male distinguished from Tree Sparrow by grey central crown, pale grey cheeks and much larger black bib. Female and immature confusable with a variety of buntings and finches, but note combination of heavy, conical bill, near-plain greyish underparts and lack of white or yellow in tail.

TREE SPARROW · 14 CM · *Passer montanus*

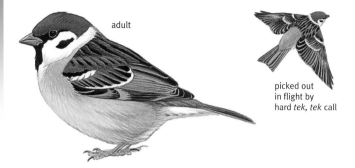

adult

picked out
in flight by
hard *tek, tek* call

juvenile

Tree
Sparrow

House
Sparrow

Smaller and daintier than House Sparrow, this country cousin has suffered an even more dramatic decline and is similarly on the conservation 'Red List'.

DESCRIPTION Crown chestnut, cheeks and collar on hindneck white, neat black bib and black spot on cheeks. Sexes similar. Juvenile as adult but head pattern duller. Often quiet and unassuming and thus easily overlooked.

POPULATION Resident. Sociable and often found in loose colonies, but extensive areas between colonies remain empty as now rare in many areas, numbers having fallen by about 90% in the last 25 years.

HABITAT Farmland with plenty of overgrown hedges and mature trees, open woodland, parkland, occasionally large gardens in villages or on edges of urban areas.

VOICE Clearly sparrow-like, many calls resemble House Sparrow, most characteristic is a hard *tek...tek...tek...* in flight.

CONFUSION SPECIES Distinguished from House Sparrow by pure white cheeks and collar, uniform chestnut cap and small black bib. Much less likely to be seen in gardens on bird feeders.

juvenile

adult

The most colourful finch and an increasingly common visitor to gardens, a flock of Goldfinches is appropriately called a 'charm'.

bold yellow wing-bars

DESCRIPTION Medium sized finch with conical, pointed, off-white bill. Sexes similar. Juvenile lacks red face, having a finely streaked greyish-white head and soft streaks on breast and flanks.

POPULATION Fairly common, has spread and increased in numbers in recent decades. Formerly heavily trapped for the cagebird trade, 'Save the Goldfinch' was one of the first campaigns of the fledgling RSPB. Over 80% of British population migrate to Europe for the winter.

HABITAT Scattered trees and shrubs, including parks and gardens. Needs open ground nearby, as principal food seeds of weeds such as dandelions, teasels and thistles, but will also take seeds of alder and birch and an increasing visitor to bird feeders, attracted above all by nyjer seeds and sunflower hearts.

often feeds on teasels

VOICE A delightful liquid twittering *teewit, tewit, titititititi....* Song an elaboration of the call.

CONFUSION SPECIES Red-faced adult unmistakable. Juvenile duller, but has black wings and tail as adult, with bold yellow wing-bar, conspicuous both in flight and at rest, white rump and white in sides to tail.

BRAMBLING · 14 CM · *Fringilla montifringilla*

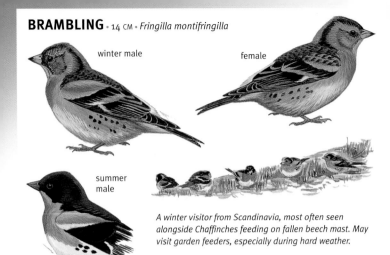

winter male

female

summer male

A winter visitor from Scandinavia, most often seen alongside Chaffinches feeding on fallen beech mast. May visit garden feeders, especially during hard weather.

Brambling, note long white rump

Chaffinch, note white outer tail feathers

DESCRIPTION Similar to Chaffinch in size and shape, but has white rump and no white in tail. Male has black head and upperparts, in autumn and winter largely hidden by brownish feather tips (and then much like female) but progressively revealed towards spring. Female and immature have mottled brown upperparts with plain greyish nape, framed by darker stripes, and greyish sides to head.

POPULATION Typically occurs in the period Oct–Apr. Bramblings wander Europe in search of beech mast, their staple winter diet, and thus although usually rather uncommon and localised in Britain, in 'irruption' years large numbers may occur. Has bred very rarely in Scotland and E England. Sociable, and almost always in flocks.

HABITAT Farmland, where feeds on stubbles and weeds, and woodland, especially beech woods.

VOICE Very harsh, nasal *arr-up*. Song a slowly repeated *dzeeeee* (very like Greenfinch's call), heard in spring before males return to boreal breeding grounds.

CONFUSION SPECIES Recalls Chaffinch, but note orange breast, shoulder and wing-bars, contrasting white belly, long white rump and, in winter, yellow bill.

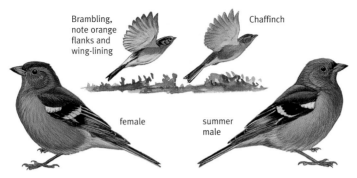

Fringilla coelebs • 14.5 CM • **CHAFFINCH**

Brambling, note orange flanks and wing-lining

Chaffinch

female

summer male

One of the commonest and most familiar birds in Britain, its cheery song is one of the easiest to learn as it is stereotyped and varies little between individuals.

DESCRIPTION Sparrow-sized, with two broad white wing-bars and white outer tail feathers. Rump greenish. Male has face and underparts pinkish-buff and crown and nape blue-grey, overall duller in winter. Female and immature much drabber, cold brown above, including cap and nape, wing-bars washed buff, underparts greyish-white.

winter male

POPULATION Common resident, with numbers augmented in winter by immigrants from Europe.

HABITAT Found anywhere where there are trees and large bushes – gardens, parks, farmland and woodland. In winter often in flocks, feeding on the ground and flying up into hedges when disturbed. A regular visitor to garden feeders.

VOICE Calls include a distinctive *pink* and loud *heap*. Song stereotyped, varying only slightly from place to place or between individuals, vigorous and endlessly repeated, a descending, staccato series of notes that terminates with a loud flourish *tink-tink-tink-chink-chink-chink-dink-dink-dink-tiddle-you-wee-o*.

CONFUSION SPECIES Female and immature are sparrow-like, but plain upperparts, pale wing-bars, pale outer tail feathers and greenish rump distinctive.

GREENFINCH · 15 CM · *Carduelis chloris*

male

female

juvenile

yellowish flashes
in wings and tail

Common in parks and gardens, and a regular visitor to feeders, this is one of the most familiar British birds. Calls and sings from prominent perches.

DESCRIPTION Chunky, sparrow-sized finch with heavy, deep-based, dull pinkish bill and yellow flashes in wings and sides of tail. Male has unmarked greenish upperparts and yellow-green underparts, female browner above, subtly streaked, and greyer below. Juvenile has heavy smudgy streaks on underparts.

POPULATION Common, although it has declined somewhat in farmland habitats. Essentially resident, but some birds leave Britain in winter while numbers are simultaneously augmented by visitors from Europe.

HABITAT Well-wooded farmland, parks and gardens, where has liking for cypress-type conifers. In winter feeds with other finches and sparrows on farmland stubble and weeds.

VOICE Very vocal, call a mellow, twittering *twi-twi-twi-twi*. Song a loud, nasal *de'weeez*, interspersed with twittering calls, sometimes given in slow, stiff-winged circular display flight.

CONFUSION SPECIES Heavy bill, more or less plain greenish appearance and flash of yellow on wings and tail distinctive.

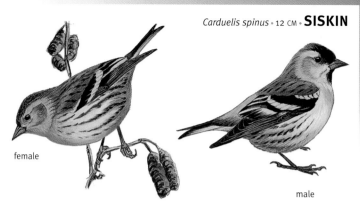

Carduelis spinus · 12 CM · **SISKIN**

female

male

This smart finch is a regular visitor to garden bird feeders, especially in late winter and early spring, with a fondness for peanuts.

DESCRIPTION Small and greenish with white, finely streaked belly and flanks and conical pointed bill. Wings black with bold yellow markings. Male has black face and bib, yellow eyebrow, breast and rump, and yellow sides to tail. Female lacks black on head, has less yellow in tail and is overall duller and more heavily streaked (including rump). Immature even duller and streakier.

POPULATION Fairly common breeding resident, most numerous in Scotland and Wales, having increased significantly in recent decades as conifer plantations mature. Numbers supplemented in winter by birds from Europe, sometimes in large 'irruptions'. Sociable and often found in flocks.

HABITAT Breeds in coniferous woodland, especially spruce plantations. Winters in stands of alder, less often birch or larch.

VOICE Flight call distinctive, high-pitched, nasal *tuwee... spiyu...tuwee...spiyu....* Flocks produce a twittering chorus. Song jangling with nasal *dweeez* notes.

CONFUSION SPECIES Smaller and neater than Greenfinch, with thinner, more pointed bill. Male distinctive, note black wing markings and finely streaked underparts of female and immature.

yellowish wing-bars, rump and tailsides (duller in female)

CROSSBILL · 16.5 CM · *Loxia curvirostra*

male

1st-year male

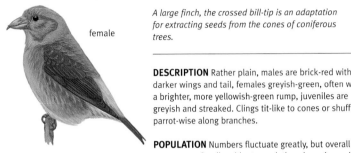

juvenile

female

female has yelowish-green rump

A large finch, the crossed bill-tip is an adaptation for extracting seeds from the cones of coniferous trees.

DESCRIPTION Rather plain, males are brick-red with darker wings and tail, females greyish-green, often with a brighter, more yellowish-green rump, juveniles are greyish and streaked. Clings tit-like to cones or shuffles parrot-wise along branches.

POPULATION Numbers fluctuate greatly, but overall uncommon. Small resident populations in regions with extensive mature conifer plantations are periodically supplemented by birds from Europe. In such 'invasion' years immigrants typically arrive mid–late summer and may disperse very widely. Many remain to breed in Dec–Jan, but numbers decline slowly until the next invasion.

HABITAT Coniferous woodland, especially pine, spruce and larch. Crossbills regularly visit ponds and puddles to drink.

VOICE Often elusive in tree tops, best detected by hard, clipped *chip, chip, chip...* given both at rest and in flight; flocks give an excited, noisy chorus.

CONFUSION SPECIES Distinctive, but in Scottish Highlands both Scottish Crossbill *L. scotica* and Parrot Crossbill *L. pytyopsittacus* breed; both are a little heavier-billed but almost impossible to separate with certainty.

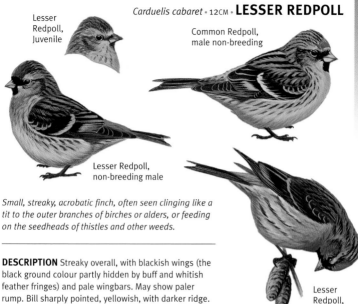

Carduelis cabaret • 12CM • **LESSER REDPOLL**

Lesser Redpoll, Juvenile

Common Redpoll, male non-breeding

Lesser Redpoll, non-breeding male

Small, streaky, acrobatic finch, often seen clinging like a tit to the outer branches of birches or alders, or feeding on the seedheads of thistles and other weeds.

DESCRIPTION Streaky overall, with blackish wings (the black ground colour partly hidden by buff and whitish feather fringes) and pale wingbars. May show paler rump. Bill sharply pointed, yellowish, with darker ridge. Face pattern distinctive, with narrow black mask, black chin and red forehead. Breeding males show pinkish on throat and breast.

Lesser Redpoll, male breeding

POPULATION Fairly common in some areas, but has undergone substantial decline and scarce in much of Midlands and S England. On the conservation 'Red List'.

HABITAT Woodland, especially coniferous, and scrubby areas with plenty of willow and birch. In winter more widespread, visiting riverside trees and weedy fields. Only occasionally visits bird tables.

VOICE Flight call a mechanical, *chett-chett-chett*. Also gives a rather Greenfinch-like *jeee*. Song, usually given in flight, is a mix of the flight call and a buzzing *brrrrr*.

CONFUSION SPECIES Common (or Mealy) Redpoll *C. flammea*, an uncommon winter visitor, is subtly larger and paler, with whiter wingbars and paler rump, but conclusive identification is difficult. In summer, male Linnets also have red forehead and breast, but are much plainer, with unstreaked upperparts.

TWITE · 14 CM · *Carduelis flavirostris*

winter adult

adult

summer adult

This rather drab, easy to overlook finch has a distinctive nasal call, which is the origin of its name. Has declined recently and now on conservation 'Red List'.

DESCRIPTION Upperparts cinnamon-brown, heavily streaked blackish; wingbar buff. Underparts cinnamon-buff, grading to whitish on belly, with smudgy streaks on flanks. Males have pinkish rump, most obvious in breeding season. Bill grey-brown in summer, yellow in winter.

POPULATION Fairly common but local breeder in Scotland, especially N and W. Rare in N Ireland and N Wales, and in England around 200 pairs in S Pennines. In winter numbers boosted by immigrants from Scandinavia.

HABITAT Favours a mosaic of moorland and farmland during breeding season, especially traditional crofts. In winter mostly found on saltmarshes along E coast, where often forms large, volatile flocks.

VOICE Song a mixture of trills, buzzes and twitters. Some calls like Linnet's, but also gives distinctive nasal *tzwee*.

CONFUSION SPECIES Needs to be carefully distinguished from female and immature Linnet. Twite has bolder wingbar and is overall more cinnamon-buff in tone, with an unmarked mustard-coloured throat (whitish, with fine darker streaks, in Linnets). They never show pink on breast (conversely, Linnet never shows pinkish on rump). In winter yellow bill distinctive.

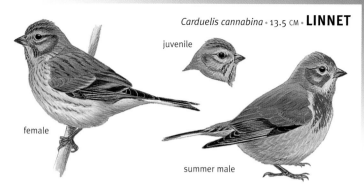

Carduelis cannabina • 13.5 CM • **LINNET**

juvenile

female

summer male

The cheery song and strawberry red breast of a male Linnet are perfectly complemented by the yellow flowers of gorse, a favoured breeding habitat.

DESCRIPTION Medium-sized finch with stubby grey bill, rather plain face with diffused paler broken eye-ring and cheek spot, white patch on closed wing and white in tail sides. In summer male has smudgy red breast and forehead and reddish-brown back. Female, winter male and immature much drabber, with brownish wash on underparts and fine streaks on crown, breast and flanks.

POPULATION Fairly common, especially in coastal areas, but has declined in recent decades due to the loss of suitable nest sites and decline of arable weeds. Now on conservation 'Red List'. Basically resident but abandons upland areas in winter and some British birds migrant to S Europe.

greyish white flashes in flight feathers and tail sides

HABITAT In breeding season favours overgrown hedges and patches of dense scrub on heaths and downland. In winter prefers farmland. Seldom visits gardens.

VOICE Flight call, especially prominent on rising, a dry, slightly nasal *chit-it, chit-it, chit-it-it*. Song a rapid trilling and jangling.

CONFUSION SPECIES Breeding male distinctive, otherwise nondescript, but note grey bill, grey head contrasting with plain brownish upperparts and white in wing.

HAWFINCH · 18 CM · *Coccothraustes coccothraustes*

Britain's largest finch, the massive bill well-adapted to cracking cherry stones. Stunning if seen well, but shy and often difficult to observe, especially in summer.

much white in wing

DESCRIPTION Bulky, short-tailed finch with large head. Bill steel-grey in summer and yellowish-brown in winter. Brown above, the cinnamon head with a narrow black mask and black bib. Underparts dull buff. Wings dark blue with a whitish shoulder. White tip to tail and white wing flash conspicuous in flight. Sexes very similar, but female has grey panel on secondaries.

POPULATION Scarce and local. The strongholds are in SE Wales, the Welsh borders, the New Forest and Home Counties. Has declined significantly in recent years and on the conservation 'Red List', but hard to find and population estimates vague.

HABITAT Deciduous woodland, usually remaining hidden in the canopy of tall trees in the summer. May descend to ground to feed in winter, which is the best time to look for them.

VOICE Call an abrupt, clicking *tsik!* (recalls a Robin's *tic*, but harder and more explosive). Song quiet and seldom heard.

CONFUSION SPECIES Waxwing has similar colouration and is only slightly larger, but is obviously crested, with yellow tips to tail and wing feathers. It is also usually much more confiding.

Pyrrhula pyrrhula • 15.5 CM • **BULLFINCH**

juvenile

female

male

This beautiful but unassuming finch is fond of areas of dense scrub and is easy to overlook unless you recognise its soft, piping whistle.

DESCRIPTION Large dumpy finch with stubby black bill, grey back and broad white wing-bars and white rump contrasting with black wings and tail. Male has rose-pink cheeks and underparts, female drab grey below, juvenile even duller and lacks black cap.

white rump conspicuous in flight

POPULATION Resident. Fairly common but, once persecuted as a pest of fruit orchards, has declined again in recent decades, probably due to loss of hedgerows, and now on conservation 'Red List'. Usually in pairs or family parties.

HABITAT Woodland, otherwise anywhere with dense thickets, such as mature hedgerows or scrub along old railways. Visits parks and gardens, especially in spring, to feed on buds. Often very skulking and difficult to see, staying in thick cover.

VOICE A beautiful soft, piping, whistle, *puh, puh....* Song soft, hesitant and seldom heard.

CONFUSION SPECIES Coloration and squat, neckless appearance distinctive, although in a brief view could be mistaken for Chaffinch.

CIRL BUNTING · 15.5 CM · *Emberiza cirlus*

summer male

female

winter male

greenish rump

male

Widespread in S England until the 1970s, but then declined rapidly to near-extinction with just 118 pairs in 1989. Numbers have since recovered.

DESCRIPTION Sparrow-sized with a dull greenish-grey rump and white outer tail feathers. Summer male has bold black and yellow head pattern and yellowish underparts with a broad greenish-grey breast band. In winter head pattern more subdued. Female much duller, with just hint of yellow on underparts. Immature duller still, finely streaked below, without yellow.

POPULATION Confined to S Devon, where around 850 pairs. Also an on-going re-introduction programme in W Cornwall (28 pairs in 2011). On the conservation 'Red List'.

HABITAT Farmland with plenty of well-grown hedges, often near the coast. Mostly feeds on stubble in winter.

VOICE Song a dry, rattling, trill lasting around 1.5 seconds, with the notes at the same pitch: *tzi-tzi-tzi-tzi-tzi....* Recalls a speeded-up Yellowhammer, but lacks the final wheezy *bzuuu*. Males often sit on prominent perches to sing. Call a thin, rather quiet *tsit*.

CONFUSION SPECIES Male distinctive, even in winter; separated from Yellowhammer by the dark throat. Female very like Yellowhammer, but head slightly more boldly-marked and, most importantly, rump dull greenish-grey, finely streaked.

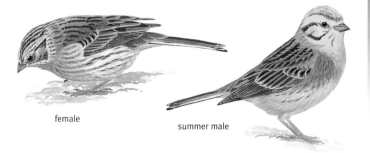

female

summer male

winter male

A classic farmland bird, the monotonous song, 'little bit of bread but no cheese' is the epitome of high summer. Like many farmland birds, has declined in recent years.

DESCRIPTION Sparrow-sized, relatively long-tailed bunting with greyish bill and rusty rump. Summer male has yellow head and underparts, smudgy rufous breast-band and finely streaked flanks. In winter yellow areas largely hidden by duller feather tips but yellow still visible around face and throat. Female similar to winter male but even less yellow, immature duller still, with no obvious yellow in plumage.

note rusty rump and white sides to tail

POPULATION Fairly common, but on conservation 'Red List' due to a rapid decline since the 1980s. Often in flocks in winter.

HABITAT Farmland, especially mixed and arable, heathland and scrub. An infrequent visitor to garden feeders, mostly to rural gardens during hard weather.

VOICE Song a stereotyped phrase, similar in all birds, 5–8 accelerating short notes followed after a gap by a longer, more wheezy note, *si-si-si-si-si—bzuuuu*. Call at rest an abrupt, metallic *tzikh*, in flight a soft *tillip*.

CONFUSION SPECIES In all plumages combination of bill shape, white-edged tail and rusty rump distinctive. Adult male could be confused with Yellow Wagtail, but note habits and rump colour. See Cirl Bunting.

SNOW BUNTING · 16.5 CM · *Plectrophenax nivalis*

winter male

winter female

summer male

summer female

A beautiful bunting, usually seen in flocks, flitting like a swirl of snowflakes along an E coast shoreline or moving like a living carpet over the ground.

DESCRIPTION Plumages complex, as male and female differ between themselves and from summer to winter, immature resembles winter female. In all plumages has white rump and outer tail and white in wing, very extensive in male.

POPULATION Scarce in winter, mostly on North Sea coasts and Scottish islands, with small numbers on coast of NW England. 70–100 pairs breed in Scotland. Very rare inland away from Highlands, but occasional migrants seen on high hills.

HABITAT Winters (Oct–Mar) on sand or shingle shores or the edge of saltmarshes, feeding on weed seeds and debris along tideline, also winters a lower altitudes in Scottish mountains; often tame. Breeds in tundra-like vegetation on high mountain tops, usually near persistent snow patches.

VOICE Soft, 'wooden', rippling *tri-li-li-lip* and more pronounced *teu*. Song a pretty warble.

CONFUSION SPECIES Extensive white in wings and tail makes identification straightforward, but beware partial albino individuals of commoner species (e.g. Linnet, House Sparrow).

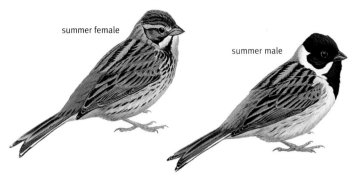

summer female

summer male

winter male

Most conspicuous in spring and summer, when males give their simple, repetitive song from the top of a prominent reed or bush.

DESCRIPTION Medium-sized bunting with small dark bill. In summer male has black hood with white moustache, collar and underparts, in winter black largely concealed, but still has vague dark bib. Female and immature brown streaked blackish above, off-white or buffish below with smudgy streaks. White outer tail feathers prominent in flight.

POPULATION Fairly common resident, but declined rapidly in the 1970s and 1980s and on conservation 'Red List'. Some immigration from the Continent in winter.

HABITAT Rank vegetation of reeds, sedges, nettles, meadowsweet, etc. on damp or wet ground, usually near water, less commonly young forestry plantations and rape fields. Winters in fields and on waste ground. A scarce visitor to garden bird feeders in rural areas, especially during cold snaps.

VOICE Song a slow series of notes, e.g. *chip, chip, siu, sirrr.* Call a thin, down-slurred *psiu*.

CONFUSION SPECIES Black hood of summer male distinctive, otherwise sparrow-like, but note bold buff eyebrow, pale moustache and streaked underparts. More strongly patterned than female and immature Yellowhammer, with dull greyish rump.

greyish rump and white sides to tail

CORN BUNTING · 18 CM · *Emberiza calandra*

adult

streaks on breast may form dark patch

heavy head and bill and pink legs are conspicuous

One of the most nondescript of British birds, most obvious when giving its cheerful jangling song from a prominent perch.

sometimes flies with legs dangling

DESCRIPTION A chunky, large-billed bunting, streaked both above and below, with streaks on breast sometimes merging into dark blotch. Note pale eye-ring and pinkish-yellow bill-base. Sexes and ages similar.

POPULATION Rather uncommon and local. Has declined sharply in recent decades, probably due to lack of winter food; now on the conservation 'Red List' and almost extinct in Ireland. Found in flocks in winter, when often elusive, but can sometimes be seen at large communal roosts. Resident.

HABITAT Extensive open farmland, particularly cereal cultivation. Sings from bushes, wires or posts. Rarely visits gardens.

VOICE Song short and stereotyped, almost identical in all individuals and said to resemble the vigorous jangling of a bunch of keys; starts hesitantly and then accelerates to a jangling flourish: *tit, tit, tiddle-idle-ee-e*. Calls include a soft *pit*.

CONFUSION SPECIES Separated from Skylark by heavy bill, and from it and other buntings, especially female Yellowhammer, by all-dark tail. From female House Sparrow by larger size, more evenly marked upperparts and coarsely streaked underparts.

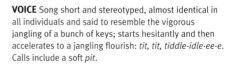

INDEX